AF597362

CRC Series in Chromatography

Editors-in-Chief

Gunter Zweig, Ph.D. and Joseph Sherma, Ph.D.

General Data and Principles
Gunter Zweig, Ph.D. and
Joseph Sherma, Ph.D.

Lipids
Helmut K. Mangold, Dr. rer. nat.

Hydrocarbons
Walter L. Zielinski, Jr., Ph.D.

Carbohydrates
Shirley C. Churms, Ph.D.

Inorganics
M. Qureshi, Ph.D.

Drugs
Ram Gupta, Ph.D.

Phenols and Organic Acids
Toshihiko Hanai, Ph.D.

Terpenoids
Carmine J. Coscia, Ph.D.

Amino Acids and Amines
S. Blackburn, Ph.D.

Steroids
Joseph C. Touchstone, Ph.D.

Polymers
Charles G. Smith,
Norman E. Skelly, Ph.D.,
Carl D. Chow, and Richard A. Solomon

Pesticides and Related Organic Chemicals
Joseph Sherma, Ph.D. and
Joanne Follweiler, Ph.D.

Plant Pigments
Hans-Peter Köst, Ph.D.

Nucleic Acids and Related Compounds
Ante M. Krstulovic, Ph.D.

CRC Handbook of Chromatography

Nucleic Acids and Related Compounds

Volume I
Part B

Editor

Ante M. Krstulović, Ph.D.
Recherche Analytique et Controle Pharmaceutique
Laboratoires D'Etudes et de Recherches Synthélabo
Paris, France

Editors-in-Chief

Gunter Zweig, Ph.D.
President
Zweig Associates
Arlington, Va.
(Deceased)

Joseph Sherma, Ph.D.
Professor of Chemistry
Lafayette College
Easton, Pennsylvania

CRC Press, Inc.
Boca Raton, Florida

Library of Congress Cataloging-in-Publication Data

Nucleic acids and related compounds.

(CRC handbook of chromatography)
Includes bibliographies and index.
1. Nucleic acids—Analysis—Handbooks, manuals, etc.
2. Chromatographic analysis—Handbooks, manuals, etc.
I. Krstulovic, Ante M. II. Zweig, Gunter. III. Sherma,
Joseph. IV. Series. [DNLM: 1. Chromatography—methods.
2. Nucleic Acids—analysis. QU 58 N9652]
QP620.N796 1987 547.7′9046 86-9603
ISBN 0-8493-3019-X (Set)
ISBN 0-8493-3020-3 (v. 1, part A)
ISBN 0-8493-3021-1 (v. 1, part B)

This book represents information obtained from authentic and highly regarded sources. Reprinted material is quoted with permission, and sources are indicated. A wide variety of references are listed. Every reasonable effort has been made to give reliable data and information, but the author and the publisher cannot assume responsibility for the validity of all materials or for the consequences of their use.

All rights reserved. This book, or any parts thereof, may not be reproduced in any form without written consent from the publisher.

Direct all inquiries to CRC Press, Inc., 2000 Corporate Blvd., N.W., Boca Raton, Florida, 33431.

© 1987 by CRC Press, Inc.

International Standard Book Number 0-8493-3019-X (Set)
International Standard Book Number 0-8493-3020-3 (Volume I, Part A)
International Standard Book Number 0-8493-3021-1 (Volume I, Part B)

Library of Congress Card Number 86-9603
Printed in the United States

VOLUME PREFACE

Since the initial publication of Volumes I and II of the *Handbooks of Chromatography* in 1972, chromatographic techniques for the separation, identification, and quantitative analysis of many classes of compounds have been the subject of thousands of additional articles. We decided, therefore, that instead of attempting to revise and update the original volumes of this series, we would publish separate volumes devoted to specific classes of chemicals. Thus, the present volumes covering the chromatography of nucleic acids and related compounds, assembled by the editor Dr. Ante M. Krstulović, represents another significant addition in this continuing series.

Various techniques of HPLC are most widely used for the separation of nucleic acids and related compounds. Therefore, the format of this volume differs from that of previous volumes of the *Handbooks of Chromatography* by not being solely organized according to chromatographic techniques but according to chemical classifications. Dr. Krstulović has obtained contributions from international experts in the separation of specific classes of compounds. We believe that this approach makes this volume most useful to readers interested in nucleic acids and related compounds. We invite comments regarding the contents and organization of this book so that future handbooks may be modified to reflect the users' preference.

Other volumes in this expanded series that have been published previously cover the following classes of chemicals: drugs (2 volumes), lipids (2 volumes), amino acids and amines, terpenoids, pesticides, carbohydrates, phenols and organic acids, polymers, and steroids. Other volumes in preparation include pigments, hydrocarbons by gas chromatography, inorganics; peptides; and Volume III of drugs, Volume II of pesticides, and Volume II of amino acids and amines.

As always, we invite our devoted readers and users of these handbooks to communicate with us as series editors or directly with volume editors with specific comments and corrections or additions. We also invite our colleagues who utilize chromatography in their research to suggest other topics and possible contributors for future *Handbooks of Chromatography*.

Gunter Zweig, Ph.D.
Joseph Sherma, Ph.D.

THE EDITORS-IN-CHIEF

Gunter Zweig, Ph.D., received his undergraduate training at the University of Maryland, College Park, where he was awarded the Ph.D. in biochemistry in 1952. Two years following his graduation, Dr. Zweig was affiliated with the late R. J. Block, pioneer in paper chromatography of amino acids. Zweig, Block, and Le Strange wrote one of the first books on paper chromatography which was published in 1952 by Academic Press and went into three editions, the last one authored by Gunter Zweig and Dr. Joe Sherma, the co-Editor-in-Chief of this series. *Paper Chromatography* (1952) was also translated into Russian.

From 1953 to 1957, Dr. Zweig was research biochemist at the C. F. Kettering Foundation, Antioch College, Yellow Springs, Ohio, where he pursued research on the path of carbon and sulfur in plants, using the then newly developed techniques of autoradiography and paper chromatography. From 1957 to 1965, Dr. Zweig served as lecturer and chemist, University of California, Davis and worked on analytical methods for pesticide residues, mainly by chromatographic techniques. In 1965, Dr. Zweig became Director of Life Sciences, Syracuse University Research Corporation, New York (research on environmental pollution), and in 1973 he became Chief, Environmental Fate Branch, Environmental Protection Agency (EPA) in Washington, D.C. From 1980 to 1984 Dr. Zweig was Visiting Research Chemist in the School of Public Health, University of California, Berkeley, where he was doing research on farmworker safety as related to pesticide exposure.

During his government career, Dr. Zweig continued his scientific writing and editing. Among his works are (many in collaboration with Dr. Sherma) the now 11-volume series on *Analytical Methods for Pesticides and Plant Growth Regulators* (published by Academic Press); the pesticide book series for CRC Press; co-editor of *Journal of Toxicology and Environmental Health;* co-author of basic review on paper and thin-layer chromatography for *Analytical Chemistry* from 1968 to 1980; co-author of applied chromatography review on pesticide analysis for *Analytical Chemistry,* beginning in 1981.

Among the scientific honors awarded to Dr. Zweig during his distinguished career are the Wiley Award in 1977, Rothschild Fellowship to the Weizmann Institute in 1963/64; the Bronze Medal by the EPA in 1980.

Dr. Zweig has authored or co-authored over 80 scientific papers on diverse subjects in chromatography and biochemistry, besides being the holder of three U.S. patents. In 1985, Dr. Zweig became president of Zweig Associates, Consultants in Arlington, Va.

Joseph Sherma, Ph.D., received a B.S. in Chemistry from Upsala College, East Orange, N.J., in 1955 and a Ph.D. in Analytical Chemistry from Rutgers University in 1958. His thesis research in ion exchange chromatography was under the direction of the late William Rieman, III. Dr. Sherma joined the faculty of Lafayette College in September 1958, and is presently Charles A. Dana Professor and Head of the Chemistry Department. He is in charge of two courses in analytical chemistry, quantitative analysis and instrumental analysis. At Lafayette he has continued research in chromatography and had additionally worked a total of 14 summers in the field with Harold Strain at the Argonne National Laboratory, James Fritz at Iowa State University, Gunter Zweig at Syracuse University Research Corporation, Joseph Touchstone at the Hospital of the University of Pennsylvania, Brian Bidlingmeyer at Waters Associates, and Thomas Beesley at Whatman, Inc. and Advanced Separation Technologies, Inc.

Dr. Sherma and Dr. Zweig co-authored or co-edited the original Volumes I and II of the *CRC Handbook of Chromatography,* a book on paper chromatography, seven volumes of the series *Analytical Methods for Pesticides and Plant Growth Regulators,* and the Handbooks of Chromatography of drugs, carbohydrates, polymers, phenols and organic acids, amino acids and amines, pesticides, terpenoids, and lipids. Other books in the pesticide series and further volumes of the *CRC Handbook of Chromatography* are being edited with Dr. Zweig, and Dr. Sherma has co-authored the handbook on pesticide chromatography. Books on quantitative TLC and advances in TLC were edited jointly with Dr. Touchstone. A general book on TLC was written with Dr. Bernard Fried, the second edition of which is now being written. Dr. Sherma has been co-author of nine biennial reviews of column and thin layer chromatography (1968—1984) and the 1981, 1983, and 1985 reviews of pesticide analysis for the ACS journal *Analytical Chemistry*.

Dr. Sherma has written major invited chapters and review papers on chromatography and pesticides in *Chromatographic Reviews* (analysis of fungicides), *Advances in Chromatography* (analysis of nonpesticide pollutants), Heftmann's *Chromatography* (chromatography of pesticides), Race's *Laboratory Medicine* (chromatography in clinical analysis), *Food Analysis: Principles and Techniques* (TLC for food analysis), *Treatise on Analytical Chemistry* (paper and thin layer chromatography), *CRC Critical Reviews in Analytical Chemistry* (pesticide residue analysis), *Comprehensive Biochemistry* (flat bed techniques), *Inorganic Chromatographic Analysis* (thin layer chromatography), *Journal of Liquid Chromatography* (advances in quantitative pesticide TLC), and *Preparative Liquid Chromatography* (strategy of preparative TLC). Dr. Sherma is editor for residues and elements of the *AOAC* and is scientific coordinator of the AOAC.

Dr. Sherma spent 6 months in 1972 on sabbatical leave at the EPA Perrine Primate Laboratory, Perrine, Fla., with Dr. T. M. Shafik, two summers (1975, 1976) at the USDA in Beltsville, Md. with Melvin Getz doing research on pesticide residue analysis methods development, and one summer (1984) in the food safety research laboratory of the Eastern Regional Research Center, Philadelphia, with Daniel Schwartz doing research on the isolation and analysis of mutagens in cooked meat. He spent three months in 1979 on sabbatical leave with Dr. Touchstone developing clinical analytical methods. A total of more than 270 papers, books, book chapters, and oral presentations concerned with column, paper, and thin layer chromatography of metal ions, plant pigments, and other organic and biological compounds; the chromatographic analysis of pesticides; and the history of chromatography have been authored by Dr. Sherma, many in collaboration with various co-workers and students. His major research area at Lafayette is currently quantitative TLC (densitometry), applied mainly to clinical analysis and pesticide residue and food additive determinations.

Dr. Sherma has written an analytical quality control manual for pesticide analysis under contract with the USEPA and has revised this and the EPA Pesticide Analytical Methods Manual under a 4-year contract jointly with Dr. M. Beroza of the AOAC. Dr. Sherma has also written an instrumental analysis quality assurance manual and other analytical reports for the U.S. Consumer Product Safety Commission, and a manual on the analysis of food additives for the FDA, both of these projects as technical editor for the AOAC. He is preparing three additional FDA manuals on animal drug and food additives analysis, and analytical field operations.

Dr. Sherma taught the first prototype short course on pesticide analysis, with Henry Enos of the EPA, for the Center for Professional Advancement. He was editor of the Kontes TLC quarterly newsletter for 6 years and also has taught short courses on TLC for Knotes and the Center for Professional Advancement. He is a consultant for numerous industrial companies and federal agencies on chemical analysis and chromatography and regularly referees papers for analytical journals and research proposals for government agencies. At Lafayette, Dr. Sherma, in addition to analytical chemistry, teaches general chemistry and a course in thin layer chromatography.

Dr. Sherma has received two awards for superior teaching at Lafayette College and the 1979 Distinguished Alumnus Award from Upsala College for outstanding achievements as an educator, researcher, author and editor. He is a member of the ACS, Sigma Xi, Phi Lambda Upsilon, SAS, AIC, and AOAC.

THE EDITOR

Dr. Ante M. Krstulović graduated in Chemical Engineering from the University of Zagreb, Yugoslavia in 1969. After a 2-year period during which he worked in industry in Yugoslavia, he obtained a Fulbright Scholarship to pursue his studies in the U.S. He completed his Master's Degree in 1971 in Analytical Chemistry from the Chemistry Department at the University of Rhode Island, Kingston.

In 1978 he obtained his Ph.D. degree from the same University. Between 1978 and 1981 he taught in the Chemistry Department at Manhattanville College, Purchase, N.Y. and pursued his research in the area of biological/biomedical applications of HPLC in nucleic acid research, tryptophan metabolism, and catecholamine measurements in various disease states.

Between 1981 and 1983 he worked as a postdoctoral fellow in the laboratory of Prof. Georges Guiochon at Ecole Polytechnique, Palaiseau, France. His research interest involved different thermodynamic aspects of liquid chromatography, such as study of selectivity, void volume determination, optimization of separations, etc.

He is currently working in the Laboratory of Analytical Research and Quality Control at Laboratoires d'Etudes et de Recherches Synthélabo, Paris.

Dr. Krstulović is the author of over 60 original publications dealing with various aspects of HPLC and a co-author of several books, such as *Reversed-Phase High-Performance Liquid Chromatography; Theory, Practice and Biomedical Applications,* Wiley-Interscience, 1982; *A Guide to the HPLC Literature,* Wiley-Interscience, 1984; *Applications of Liquid Chromatography to the Development of Pharmaceuticals,* Astor Publishing Corp., 1985; *Quantitative Analysis of Catecholamines and Related Compounds,* Ellis Horwood, 1986.

CONTRIBUTORS

Lon Alterman
SOTA Chromatography
Crompond, New York

Rosario Ammendola, B.S.
Researcher
Istituto di Scienze Biochimiche
University of Naples
Naples, Italy

Eberhard Amtmann
Institute for Virus Research
German Cancer Research Center
Heidelberg, West Germany

Paul A. Andrews, Ph.D.
Assistant Research Biochemist
Cancer Center
University of California, San Diego
La Jolla, California

Arthur H. Bertelsen, Ph.D.
Group Leader
Department of Molecular Biology
Unigene Laboratories, Inc.
Fairfield, New Jersey

Rainer Bischoff, Ph.D.
Department of Biochemistry
Purdue University
West Lafayette, Indiana

Sylvain Blanquet, D.Sc.
Director
Laboratoire de Biochimie
Ecole Polytechnique
Palaiseau, France

Karl-Siegfried Boos, Dr. rer. nat.
Retired
Laboratorium für Biologische Chemie
Universität Paderborn
Paderborn, West Germany

Thomas Braumann
Gamma Analysen Technik (GAT)
Bremerhaven, West Germany

Annie Brevet, D.Sc.
Research Associate
Laboratoire de Biochimie
Ecole Polytechnique
Palaiseau, France

Eric G. Brown, Ph.D., D.Sc., C.Chem., F.R.S.C.
Professor and Head
Department of Biochemistry
University College of Swansea
Swansea, Wales

Judith K. Christman, Ph.D.
Professor of Biochemistry
Research Professor of Pediatrics
Mount Sinai Medical Center
New York, New York

Filiberto Cimino, M.D.
Professor of Biochemistry
Istituto de Scienze Biochimiche
University of Naples
Naples, Italy

Aharon S. Cohen, Ph.D.
Department of Inorganic and Analytical Chemistry
Hebrew University
Jerusalem, Israel

Henri Colin, Ph.D.
Technical Director
PROCHROM S.A.
Nancy, France

Alfredo Colonna, B.S.
Associate Professor
Istituto di Scienze Biochimiche
University of Naples
Naples, Italy

Angelo P. Consalvo, B.S.
Associate Scientist
Department of Organic Chemistry
Unigene Laboratories, Inc.
Fairfield, New Jersey

Jan Willem de Jong, Ph.D.
Director
Cardiochemical Laboratory
Erasmus University
Rotterdam, Netherlands

Peter P. de Tombe
Department of Medicine and Medical Physiology
Foot Hills Hospital
University of Calgary
Calgary, Alberta, Canada

Angela Duilio, B.S.
Researcher
Istituto di Scienze Biochimiche
University of Naples
Naples, Italy

Franca Esposito, M.D.
Associated Researcher
Istituto di Scienze Biochimiche
University of Naples
Naples, Italy

Carlo Fini, Ph.D.
Associate Professor
Istituto Interfacoltà di Chimica Biologica
University of Perugia
Perugia, Italy

Ardesio Floridi, Ph.D.
Professor of Applied Biochemistry
Istituto Interfacoltà di Chimica Biologica
University of Perugia
Perugia, Italy

Peter Földi, Dr. rer. nat.
LKB Instrument GmbH
Gräfelfing, West Germany

Charles W. Gehrke, Ph.D.
Professor of Biochemistry
Director, Interdisciplinary Chromatography Mass Spectrometry Facility
University of Missouri-Columbia
Columbia, Missouri

James P. Gilligan, Ph.D.
Senior Scientist
Department of Protein Chemistry
Unigene Laboratories, Inc.
Fairfield, New Jersey

Marie F. Gonnord, Ph.D.
Senior Researcher
Department de Chemie
Ecole Polytechnique
Palaiseau, France

Lutz Graeve, Dr. rer. nat.
Department of Molecular Biology
Institute for Physiological Chemistry
University of Hamburg
Hamburg, West Germany

Eli Grushka, Ph.D.
Professor
Department of Inorganic and Analytical Chemistry
Hebrew University
Jerusalem, Israel

Norberto Guzman
Princeton Laboratories
Princeton, New Jersey

Eberhard Hagemeier, Dr. rer. nat.
Laboratorium für Biologische Chemie
Universität Paderborn
Paderborn, West Germany

R. A. Harkness, Ph.D., F.R.C.P.E.
Division of Inherited Metabolic Diseases
MRC Clinical Research Center
Harrow, Middlesex, England

Eef Harmsen, Ph.D.
Department of Biochemistry
University of Oxford
Oxford, England

Johan A. J. Hegge
Cardiochemical Laboratory
Erasmus University
Rotterdam, Netherlands

Shingo Hirose, Ph.D.
Professor
Department of Analytical Chemistry
Kyoto Pharmaceutical University
Kyoto, Japan

Masaki Iwamoto, Ph.M.
Research Worker
Biwako Research Laboratory
Ohtsuka Pharmaceutical Company
Kyoto, Japan

Barry N. Jones, Ph.D.
Group Leader
Department of Protein Chemistry
Unigene Laboratories, Inc.
Fairfield, New Jersey

Klaus Kemper, Dr. rer. nat.
Laboratorium für Biologische Chemie
Universität Paderborn
Paderborn, West Germany

Wilfried Kleiböhmer
Inorg. Chemical Institute
Westfälische Wilhelmsuniversität
Münster, West Germany

J. C. Kraak, Ph.D.
Laboratory of Analytical Chemistry
University of Amsterdam
Amsterdam, Netherlands

Peter Krieg
Institute of Virus Research
Germany Cancer Research Center
Heidelberg, West Germany

Ante M. Krstulović, Ph.D.
Recherche Analitique et Contrôle Pharmaceutique
L.E.R.S.-SYNTHELABO
Paris, France

Joachim Kruppa, Dr. rer. nat.
Professor
Department of Molecular Biology
Institute for Physiological Chemistry
University of Hamburg
Hamburg, West Germany

Kenneth C. Kuo, M.S.
Senior Research Chemist and Chromatographer
Experiment Station Chemical Laboratories
University of Missouri-Columbia
Columbia, Missouri

Leonard F. Liebes, Ph.D.
Research Assistant Professor
Division of Oncology
Department of Medicine
New York University
New York, New York

Larry W. McLaughlin
Department of Chemistry
Max-Planck Institute for Experimental Medicine
Gottingen, West Germany

Naomi Mendelsohn, Ph.D.
Assistant Professor
Department of Pediatrics
Mount Sinai Medical Center
New York, New York

Bernard Mompon, D.Sc.
Head, Structural Analysis Department
L.E.R.S.-SYNTHELABO
Bagneux, France

Erkki A. O. Nissinen, Ph.D.
Head, Biochemical Laboratory
Department of Pharmacology
Orion Pharmaceutical Company
Espoo, Finland

Carlo A. Palmerini
Research Biochemist
Istituto Interfacoltà di Chimica Biologica
University of Perugia
Perugia, Italy

David Perret
Research Biochemist
Department of Medicine
Dunn Laboratory
St. Bartholomew's Hospital Medical College
London, England

Pierre Plateau, D.Sc.
Research Associate
Groupe de Biophysique
Ecole Polytechnique
Palaiseau, France

Andreas Rizzi, Ph.D.
Institute of Analytical Chemistry
University of Vienna
Vienna, Austria

Tommaso Russo, M.D.
Associated Researcher
Istituto di Scienze Biochimiche
University of Naples
Naples, Italy

Francesco Salvatore, M.D.
Professor of Biochemistry
Istituto di Scienze Biochimiche
University of Naples
Naples, Italy

Gerhard Sauer, Ph.D.
Professor
Institute for Virus Research
German Cancer Center
Heidelberg, West Germany

Eckhard Schlimme, Dr. sc. agr., Dr. rer. nat.
Professor
Laboratorium für Biologische Chemie
Universität Paderborn
Paderborn, West Germany

Roger J. Simmonds, Ph.D.
Department Metabolism
Wellcome Research
Beckenham, Kent, England

Albert H. van Gennip, Ph.D.
Clinical Chemist
Head of Laboratories
Laboratory for Metabolic Disorders and Cancer
Children's Hospital
Amsterdam, Netherlands

Peter J. M. Van Haastert, Ph.D.
Department of Cell Biology and Genetics
University of Leiden
Leiden, Netherlands

Hans Veening, Ph.D.
Professor and Chairman
Department of Chemistry
Bucknell University
Lewisburg, Pennsylvania

Bernd Wenclawiak, Dr. rer. nat.
Associate Professor
Department of Chemistry
University of Toledo
Toledo, Ohio

Shigeru Yoshida, Ph.D.
Assistant Professor
Pharmaceutical Sciences
Kyoto Pharmaceutical University
Kyoto, Japan

Stanley D. Young, Ph.D.
Senior Scientist
Department of Organic Chemistry
Unigene Laboratories
Fairfield, New Jersey

TABLE OF CONTENTS

Part A

Part B

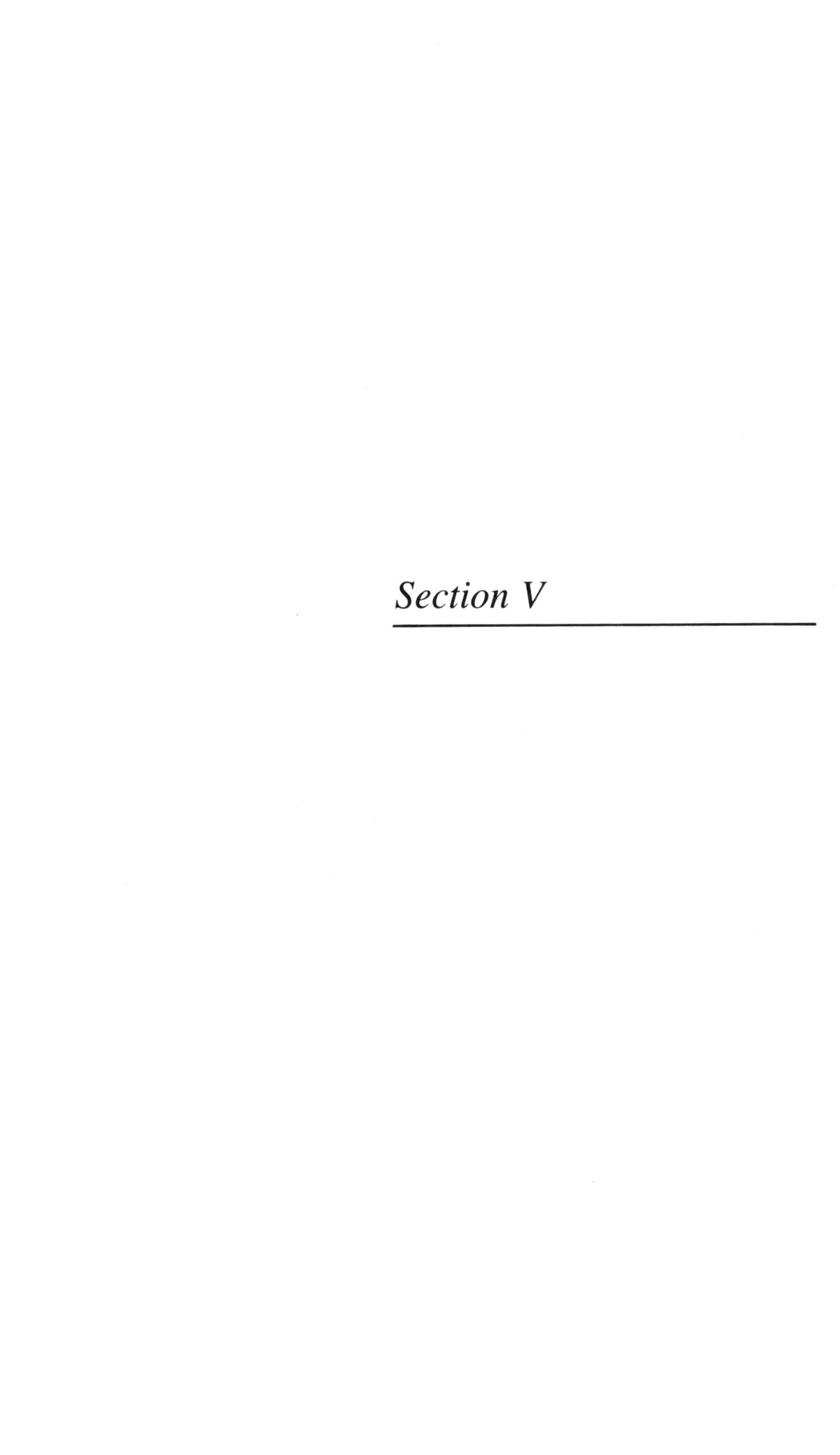

Section V

ISOLATION AND EXTRACTION OF NUCLEOTIDES FROM TISSUES PRIOR TO HIGH PERFORMANCE LIQUID CHROMATOGRAPHY

David Perrett

INTRODUCTION

Nucleotides occur in the majority of tissues from all living species at concentrations of up to 10 μmol/mℓ of cell water. Early studies often focused on the adenine nucleotides, partly because of their central role in energy metabolism and cellular regulation, and for the very practical reason that they are the most abundant and, therefore, the easiest to measure by the variety of nonchromatographic assays then available. ATP is found in substantial amounts (up to 4 μmol/g wet weight) in well-prepared tissue extracts; e.g., in liver it can form over 60% of the total nucleotide pool, while ADP and AMP form only 15 and 5%, respectively. With the application of chromatographic techniques to the separation of nucleotides, it has been possible to determine at least 30 nucleotide peaks in many tissue extracts, but many of these components occur at less than 1% of the ATP concentration. The majority of these measurable components are ribonucleotides, with the deoxynucleotides and the cyclic nucleotide pools not normally exceeding 50 and 30 nmol/g wet weight, respectively. This level of complexity, as well as the wide concentration range to be measured, would in itself present a marked challenge to the biochemical analyst, but the task is further complicated by the instability, both biochemical and chemical, of nucleotides in tissues and tissue extracts.

Nucleotides are rapidly degraded by a wide variety of intracellular enzymes, including 5′-nucleotidases to lower order nucleotides, such as AMP, and then to nucleosides and bases. These, in turn, may be salvaged and reformed into nucleotides, e.g., hypoxanthine may be converted to IMP by the action of HGPRTase. An outline of these pathways is presented in Figure 1. It is important, therefore, to block these interconversions in the time interval between sampling the tissue and extracting the nucleotides for analysis. Since this degradation is mainly biochemical in origin, its rate is tissue dependent. For example, substantial degradation of liver nucleotides occurs within seconds at 37°C, whereas human red cell nucleotides are relatively stable for several hours. The extraction techniques employed will, therefore, reflect these differences.

In order to achieve good chromatographic resolution, consideration must also be given to the nature of the actual sample to be injected. Although direct injection of unextracted material onto the HPLC column can be used for some samples, e.g., very small volumes of plasma, this is not usually practicable with crude tissue homogenates. Injection of such solutions containing high concentrations of proteins could rapidly ruin the analytical column. To increase column lifetimes, various procedures, which remove the majority of the protein while releasing the nucleotides, have been employed. With relatively nonspecific detection techniques, such as UV absorption, it is advantageous if the sample preparation procedure is selective, since this will reduce the number of possible interferences in the final chromatogram.

Any extraction technique is required to fulfill the following criteria in order to approach the ideal:

1. It should free all of the compound(s) of interest from within the tissue, including that which is protein bound.
2. It should precipitate all the protein in order to avoid detrimental effects on column performance.

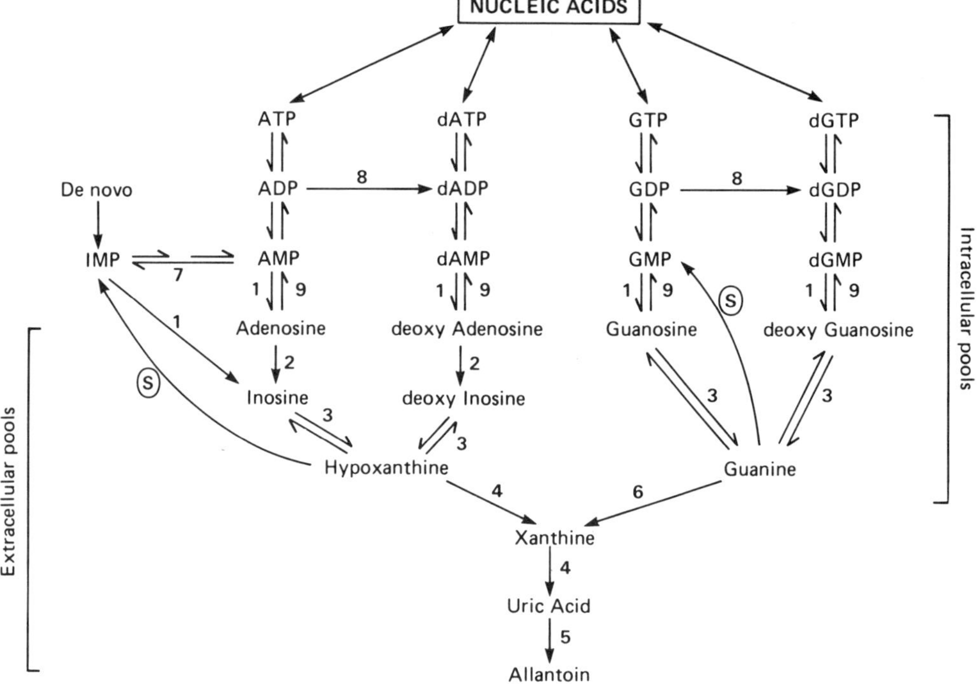

FIGURE 1. Degradative and salvage pathways for the purine nucleotides, which may result in changes to the nucleotide content of cells during warm ischemia. Note: not all pathways are shown and some of those shown do not operate in all tissue types. (1) 5′-Nucleotidase; (2) adenosine deaminase; (3) purine nucleoside phosphorylase; (4) xanthine oxidase; (5) allantoinase; (6) guanase; (7) AMP deaminase; (8) ribonucleoside-diphosphate reductase; (9) (deoxy)nucleoside kinase; (S) hypoxanthine-guanine phosribosyltransferase.

3. It should not degrade the compounds either chemically or permit biochemical degradation to continue. Equally, it should not degrade polymeric species, and thereby, increase the concentration of monomers.
4. Preferably, the extractant should not interfere with the assay or it should be easily removed prior to analysis.
5. It is advantageous if it can selectively extract the compounds of interest from others with similar chromatographic and/or detection characteristics.
6. The final extract should be clean and free of particulate matter.
7. The procedure should not dilute the extract unnecessarily.
8. The compound(s) should be stable in the final extract, particularly if it is necessary to store it prior to analysis.
9. The complete procedure should be reproducible and give good recoveries.
10. It should be simple and rapid so large numbers of samples can be processed when necessary.

This chapter will describe the procedures which have been found most satisfactory for the isolation and extraction of nucleotides prior to HPLC analysis in this laboratory over the past 12 years. Additionally, details of other recommended methods from the literature will be given.

CHOICE OF DEPROTEINATING PROCEDURE FOR NUCLEOTIDES

A wide variety of methods have been evolved to remove proteinaceous material from

Table 1
DEPROTEINATING PROCEDURES AND THEIR MECHANISMS OF ACTION

Procedure	Principle	Suitability for nucleotides
Heating	Denatures proteins	No
Cold	Denatures membranes	No
Ultrafiltration	Molecular size separation	Possibly for free NXPs
Ammonium sulfate	Salting out	Unsuitable for HPLC
$Ba_2SO_4/Zn(OH)_2$	Co-precipitation	No
Methanol and other organics	Dielectric change solubility	Helpful along with other agents
PCA	pH denaturing	Good
TCA	pH denaturing	Excellent

biological samples (for reviews see References 2 to 5). Some of the most popular methods along with their mode of action are listed in Table 1.

Most of these methods were originally developed for other types of assay and modifications to the procedures are needed to make their use compatible with both nucleotides and HPLC. Brown and Miech[6] were the first to compare extraction procedures for use with HPLC. They extracted nucleotides from human erythrocytes, followed by separation on a pellicular ion-exchanger. Two of the methods they compared have remained the basis for extracting nucleotides from mammalian cells. They were perchloric acid (PCA) extraction followed by neutralization with potassium hydroxide and removal of the potassium perchlorate precipitate, and trichloroacetic acid (TCA) followed by extraction of the excess TCA with water-saturated diethyl ether. Other methods, including direct injection of the acid extracts or neutralization of the extract with tris(hydroxymethyl)-aminomethane(TRIS), were unacceptable, since either the nucleotides were unstable on storage or injection of these concentrated solutions disturbed the chromatography. They found that TCA and PCA gave comparable results. The PCA/KOH procedure leaves a saturated (63 mmol/ℓ) solution of $KClO_4$ and more precipitate tends to form on storing the sample at $-20°C$. Losses of nucleotides of up to 10% by adsorption onto this preciptate can occur.[95] The use of a TCA procedure is therefore preferred. Although Brown later claimed this method was unsatisfactory with microparticulate packings,[7] we have experienced no such difficulties.

The various parameters governing the efficient use of TCA, such as volume-to-tissue ratio, optimum concentration, and the efficiency of protein removal, have received little attention. Two studies[8,9] have compared how efficiently a number of procedures remove proteins from plasma. Blanchard[9] determined the percentage removal of protein by measuring residual protein using the sensitive Lowry technique. He concluded that the anionic precipitants TCA, PCA, tungstic acid, and metaphosphoric acid were the most efficient. One volume of 10% (w/v) TCA or 6% (w/v) PCA precipitated >98% of the protein in 2.5 vol of plasma; the TCA solution was slightly more efficient than the PCA. Precipitation using organic solvents required much larger volumes to achieve the same efficiency. There appears to be no comparable study on removing protein from more complex materials such as red cells or tissues. In a study on muscle[10] we observed that, although extraction with ice cold 20% TCA gave good ATP/ADP ratios, some enzyme activity remained in the extract. For example, ^{14}C-AMP was used as the internal standard; ^{14}C-ADP and ^{14}C-ATP were also recovered in the column eluate. This remaining activity of the enzymes myokinase and creatine kinase could be abolished by including 20% methanol in the TCA solution, which lowered the temperature of the extraction mixture. In addition, it also acted as a deproteinating agent. North et al.[11] have also observed nucleotide catabolizing enzymes to be present in methanol extracts of cells.

Table 2
RECOVERY OF 5′-NUCLEOTIDES RECOVERED FROM RAT LIVER BY DIFFERENT EXTRACTION PROCEDURES[a]

Nucleotide	Extraction procedure[b]		
	C-A	A-F	PCA-KOH
AMP[c]	85 ± 13	402 ± 94	228 ± 64
CMP	17 ± 1	40 ± 2	32 ± 7
GMP	17 ± 1	75 ± 7	47 ± 7
UMP	51 ± 11	113 ± 15	76 ± 16
ADP	301 ± 80	537 ± 5	637 ± 83
CDP	84 ± 0	408 ± 8	127 ± 5
GDP	68 ± 8	69 ± 3	96 ± 28
UDP	76 ± 4	141 ± 1	165 ± 8
ATP	843 ± 67	1398 ± 5	1407 ± 98
CTP	132 ± 8	108 ± 8	124 ± 14
GTP	208 ± 8	253 ± 9	263 ± 9
UTP	181 ± 9	256 ± 4	265 ± 8

[a] Values are nmol/g frozen liver powder and are mean ± SEM of three extractions from three different animals.
[b] C-A = charcoal absorption of nucleotides; A-F = TCA extraction followed by Alamine®/Freon® extraction.
[c] AMP and NAD were not resolved and this is a summation.

Reproduced from Riss, T. L., Zorich, N. L., Williams, M. D., and Richardson, A., *J. Liq. Chromatogr.*, 3, 133, 1980. By courtesy of Marcel Dekker, Inc.

Most workers have removed excess TCA by multiple extractions (minimum 3 × 4 vol) with water-saturated diethyl ether or precipitated PCA with KOH. Khym[12] introduced a combined solvent extraction-cum-neutralization procedure, whereby excess acid was removed from nucleotide containing solutions by extracting with a water-insoluble amine dissolved in a water-immiscible solvent. The amine solvent pair used was 0.5 mol/ℓ Alamine® 336 dissolved in Freon®-TF which was mixed with the aqueous extract for 3 to 4 min. Chen et al.[13] replaced Alamine® with tri-*N*-octylamine and found the method to be particularly suitable provided the concentration of amine is carefully controlled and it is freshly prepared. The Freon®/amine procedure was later optimized by Van Haverbeke and Brown[14] who concluded that the TCA concentration must be at least 9% (w/v) and that the exact amount of amine to neutralize the solution is required. Use of excess amine could reduce recovery of ATP from erythrocytes by 30%. The exact amount should be determined for each sample by titration. They also observed that recoveries of added nucleotides were 40% lower from erythrocytes than from water. This conclusion was not supported by the work of Riss and colleagues,[15] who studied the efficiency of extracting nucleotides from liver and isolated hepatocytes. In their hands the recoveries of added nucleotides using TCA plus Alamine®/Freon® were approximately 90% for 12 different nucleotides and approximately 80% for a PCA-KOH procedure (Table 2). The consensus of option is, therefore, that, at least with mammalian cells, TCA is the most efficient protein precipitating agent and that there are some advantages to using an amine/immiscible solvent neutralization scheme.

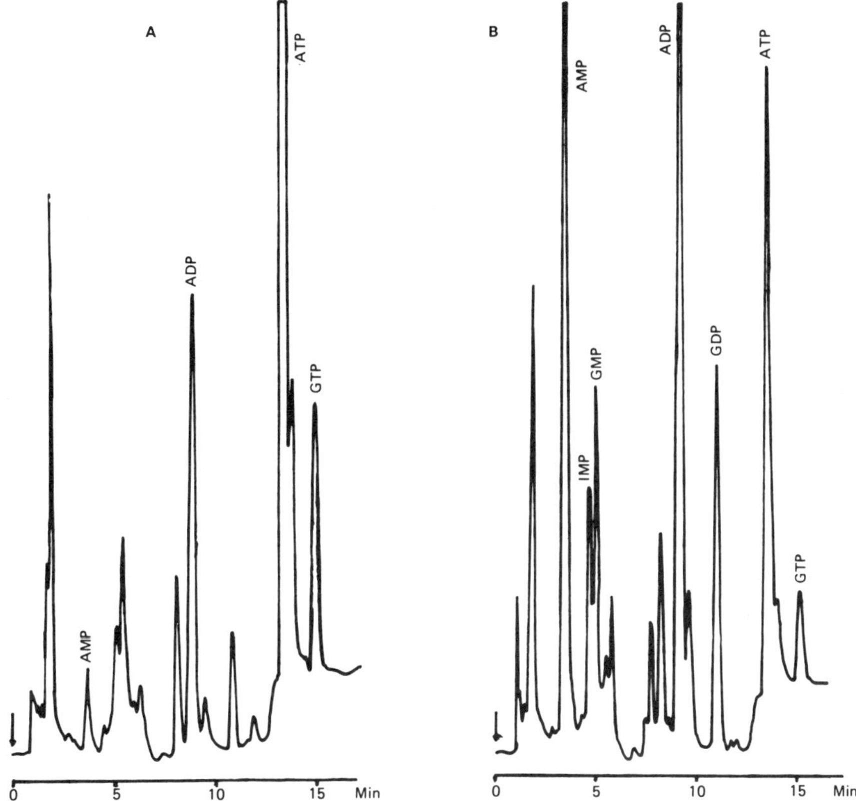

FIGURE 2. The effect of a short period of ischemia on the nucleotide pattern of rat liver. (A) Tissue immediately freeze clamped and extracted with TCA; (B) tissue extracted after 30 sec of ischemia at room temperature. Chromatograms obtained by anion-exchange gradient HPLC method of Perrett.[44] Column 100 × 0.4 mm 5 μm APS-Hypersil® Detector, CE212 254 nm 0.1 A AUFS linear phosphate gradient. A volume of extract equivalent to the same weight of tissue (3 mg) was analyzed in each case.

ISOLATION OF CELLULAR NUCLEOTIDES: SAMPLE PREPARATION

The two most-studied sources of nucleotides are tissues, particularly rat liver, and the formed elements of human blood such as red cells and platelets. The techniques necessary to achieve satisfactory preparations are a direct function of the metabolic activity of the tissue. It has been known for many years that the ATP concentration of tissues can drop by up to 100% during the dissection of liver. With the advent of HPLC analysis it has become clear that equally large changes in the concentration of other nucleotides also occur in ischemic tissues. This author is unaware of a comparative study on the rate of ischemic degradation of nucleotides by animal tissues.

Whole Tissues

In order to minimize changes in nucleotide composition due to warm ischemia, it is necessary to stop enzyme action by rapidly freezing the tissue. Freezing the sample is preferably performed *in situ* with the live animal under light anesthesia. The frozen tissue can then be removed into a cold vessel for weighing, followed by homogenization in a cold solution of protein precipitant without further changing the nucleotide content. Since the changes are time dependent, it is essential to perform all three operations in the minimum possible period of time. Faupel et al.[16] observed a 60% reduction in the ATP content of rat liver within 1 min if the sample was maintained at body heat. This was associated with a 220% increase in ADP and a 1000% increase in AMP levels. Similar increases were found for a wide variety of glycolytic intermediates such as glucose-6-P, glucose-1-P, and fructose 1,6 diP. The effect of 15 sec warm ischemia on rat liver is shown in Figure 2.

Wollenberger[17] showed that 16.3 sec are necessary to decrease the temperature of a portion of heart weighing 800 mg from 38 to 0°C by immersion in liquid nitrogen. Causes of this slow rate of freezing are the low thermal conductivity of tissue and the insulation caused by the formation of a surface film of gas. These observations were later confirmed by Faupel and colleagues,[16] who confirmed that the only practicable method of overcoming the low thermal conductivity is freeze clamping.

Freeze clamping is best performed *in situ* using metal tongs, as first described by Wollenberger.[17] The dimensions of the tongs should be appropriate to the tissue being sampled. Small, thick copper jaws are needed for sampling from mice and rats, while larger jaws are suitable for larger animals or tougher tissues. Typically, the jaws will clamp between 50 and 200 mg of wet tissue. The jaws of the tongs should be placed in liquid nitrogen (−196°C) or liquid air contained in a Dewar vessel until they attain the temperature of the bath. The handles of the tongs should be insulated with cork or similar material; the use of insulating gloves is not recommended, since these can lead to difficulties in handling and slow down the clamping process.

Ether anesthesia caused only minimal changes in the nucleotide content of those animal tissues where studies have been made.[18] Bucher and Swaffield[19] found that up to 16 min under ether anesthesia produced only slight changes to the adenosine nucleotides and ATP/ADP ratio in rat liver. In a study on muscle adenosine nucleotides of a wide variety of animals and insects, Beis and Newsholme[20] compared various forms of anesthesia. There was no significant effect on nucleotide concentrations between physical anesthesia such as a blow to the head and the use of an injection of Nembutal®, whereas significant changes were found using Sandoz®-MS222. Faupel et al.[16] found that a significant reduction in rat liver ADP and an increase in AMP occurred following 30 min anesthesia with i.p. Nembutal® (6 mg/100 g body weight) when compared to values obtained by the use of a double hatchet technique on the unanesthetized animal. No studies on the effect of anesthesia on the complete nucleotide pattern studied by HPLC has been reported. Care must be taken not to significantly traumatize the animal, since stress can also cause some degree of change.[16] Rapid light anesthesia with Nembutal® would appear to give the optimum values.

The tissue to be clamped should be exposed and surrounding skin and organs gently eased aside. The tongs should be raised from the nitrogen bath and the tissue rapidly squeezed between the flat jaws, in a single movement. In this way the tongs are dual purpose: they rapidly freeze the tissue preventing metabolic changes and, if correctly applied, i.e., with a rapid action and moderate force, they mechanically fracture the tissue. Following freeze clamping any unfrozen tissue should be crudely cut away from around the head of the tongs with a sharp, strong scalpel and the frozen pellet lifted out using the tongs. All material not trapped between the jaws must now be cut away before the pellet is eased off the jaws onto a cold weighing boat and rapidly weighed. (Some authors[15] suggest that the pellet can now be ground in a cold pestle and mortar and the powder weighed into appropriate aliquots.) The weighed frozen tissue is added to an ice-cold homogenizer tube containing an appropriate volume of ice-cold protein precipitant. Although 1 mℓ of 10% w/v TCA per 100 mg wet weight of tissue is usually sufficient, the optimum proportion should be determined for each tissue. Cold homogenization should be performed immediately and continued until the mixture appears uniform in color and texture. The time required is dependent on the mass and type of tissue and the type of homogenizer, but it is usually less than 30 sec. Both motor-driven homogenizer of the pestle and tube (Potter-Elvehjem) variety and the enclosed blade (Turax) type have been successfully employed. A microversion of the motor-driven type capable of homogenizing less than 10 mg of tissue has been described.[10] The nucleotide-containing acid supernatant can now be separated by centifugation. The excess acid should be immediately neutralized using one of the techniques described in the previous section.

In order to standardize the final results a small sample (10 to 20 mg) of freeze-clamped tissue should be collected into a preweighed tube and then dried or freeze dried to a constant weight so as to determine the wet weight/dry weight ratio of the original tissue. If required the protein or DNA concentration of the sample can be determined using standard techniques in either a sample of the final homogenate or by solubilizing the TCA precipitate.

Specific Examples

Liver and Kidney

These two tissues contain a wide variety of nucleotides and present considerable sampling and chromatographic difficulties. Substantial changes in nucleotides have been reported following warm ischemia of rat liver specimens (Figure 2). The changes are approximately linear during the first 15 sec, according to Faupel et al.[16] In human liver sampled by needle biopsy and immediately dropped into liquid Freon® (-150°C), Hultman et al.[21] could detect no additional degradation of ATP by increasing the delay between sampling and freezing from a minimum of 2 sec up to 7 sec.

Kidney extracts show a similar nucleotide profile to liver, and their extraction from fresh tissue presents similar problems to those found with liver. Ischemic changes are very rapid and freeze clamping must be performed rapidly.[22]

Heart and Skeletal Muscle

The nucleotide content of these tissues is mainly represented by the adenosine nucleotides, and even GTP is only present at relatively small concentrations. This pattern is, of course, characteristic of tissues where energy for contraction must be derived from the adenosine nucleotide pool. Freeze clamping is essential, but even then problems of completely and rapidly inhibiting enzyme activity have been encountered. These difficulties can be overcome by using TCA dissolved in methanol.[10] There have been many studies on the adenosine nucleotide pools in heart. Roberts and Bessman[23] developed methodologies to determine a complete range of sugar phosphates in fetal rat hearts. Beis and Newsholme[20] developed procedures for the accurate extraction of adenine nucleotides from the resting muscles of many species.

Intestinal Tract

In this case the difficulties are that the organ is both heterogeneous and also contains foreign matter. It is necessary to fast the animal overnight in order to empty the intestine so that the tissue can be freeze clamped, but it may also be required to wash out the colon. There is a nucleotide gradient down the length of the tissue, the majority of which is caused by the differing water content of the jejunum, ileum, and colon. There are also marked differences in the ease with which the different segments can be homogenized.

Brain

It is, of course, impossible to freeze clamp living brain tissue, but rapidly frozen tissue can be obtained by dropping small animals, e.g., rat or mouse, head first into a large volume of liquid nitrogen. The brain can then be removed using a cold chisel and homogenizing appropriate sections. Shmukler[24] demonstrated that decapitation followed by immediate freezing of the head gave an ATP/AMP ratio of 1 compared to 100 for head-first immersion.

Formed Elements of Blood

Because of the ready availability of blood cells, especially red blood cells, they have often been the model system on which extraction techniques have been developed.

Red Cells

Blood should be collected into heparin tubes and immediately centrifuged at 2000 g (or at the maximum speed of the centrifuge if slower) and preferably at 4°C. The plasma and buffy coat are then removed with a Pasteur pipette. The red cells can, if required, be washed with 1 vol of ice-cold saline. The packed red blood cells (1 vol) are slowly added to 2 vol of ice-cold 10% TCA or another protein precipitant in a plastic centrifuge tube being mixed on a vortex mixer. The resultant brown precipitate is removed by centrifugation. The supernatant containing the nucleotides is transferred to a glass test tube and the excess TCA removed as already described. An aliquot of the packed cells is retained for determination of their hematocrit, hemoglobin, or total protein concentration to which the final concentration of nucleotides will be referred.

In some cases, particularly where rapid sampling is required, it is possible to extract whole blood and the values obtained will approximate to those for red cells alone. In order to allow for the plasma water the volume and strength of the TCA used should be adjusted accordingly.

Platelets

Platelet-rich plasma (PRP) is obtained by low speed centrifugation (e.g., 500 g for 10 min) at 4°C of fresh heparinized blood. The PRP is carefully removed and the platelet count determined on an aliquot. A measured volume of the remaining PRP is centrifuged at about 2500 g for 10 min. The plasma is removed from the platelet plug and the platelets washed with a little cold saline. Following a further centrifugation, as much as possible of the washings are removed, and while vortexing, 10% TCA (100 μℓ/10 cells) is added. It is usually necessary to sonicate or homogenize the mixture to completely disintegrate the cells.

Lymphocytes

These are obtained by density sedimentation through the Ficoll/Hypaque mixture according to standard techniques. The isolated cell fraction must then be washed a minimum of three times to remove all traces of the Ficoll/Hypaque, since it contains a number of strongly UV absorbing components which interfere with the subsequent HPLC analysis (Figure 3A). The cells are then extracted in the same manner as platelets.

Plasma

Although plasma is the usual medium in which nucleosides and bases are determined, it has recently become clear that plasma also contains trace quantities of nucleotides. They may result, in part, from hemolysis of cells during collection, but by carefully controlling the collection of the blood Harkness et al.[25] have convincingly demonstrated the presence of adenylate nucleotides which are derived from other sources.

Bacteria and Tissue Culture Cells

Compared to blood cells, these cells are significantly more metabolically active and nucleotides degrade at a correspondingly faster rate during isolation.[26] The major difference between bacterial or tissue culture cells and whole tissue extraction is, therefore, cell density. Cell densities of 10^6 to 10^8 cells per milliliter are typically achieved at the end of culture periods, and cells must be harvested by either centrifugation or filtration prior to acid extraction. Payne and Ames[26] investigated both isolation procedures and concluded that centrifugation could not be performed rapidly enough to prevent ATP degradation. More consistant and reproducible results were achieved by vacuum filtration through glass fiber membranes. Up to 200 mℓ of bacterial suspension could be filtered in less than 1 min. Nucleotides were immediately extracted either *in situ* or by placing the filter in 1 mol/ℓ formic acid for 30 min. The percent of cells trapped by the filter (usually about 90%) was

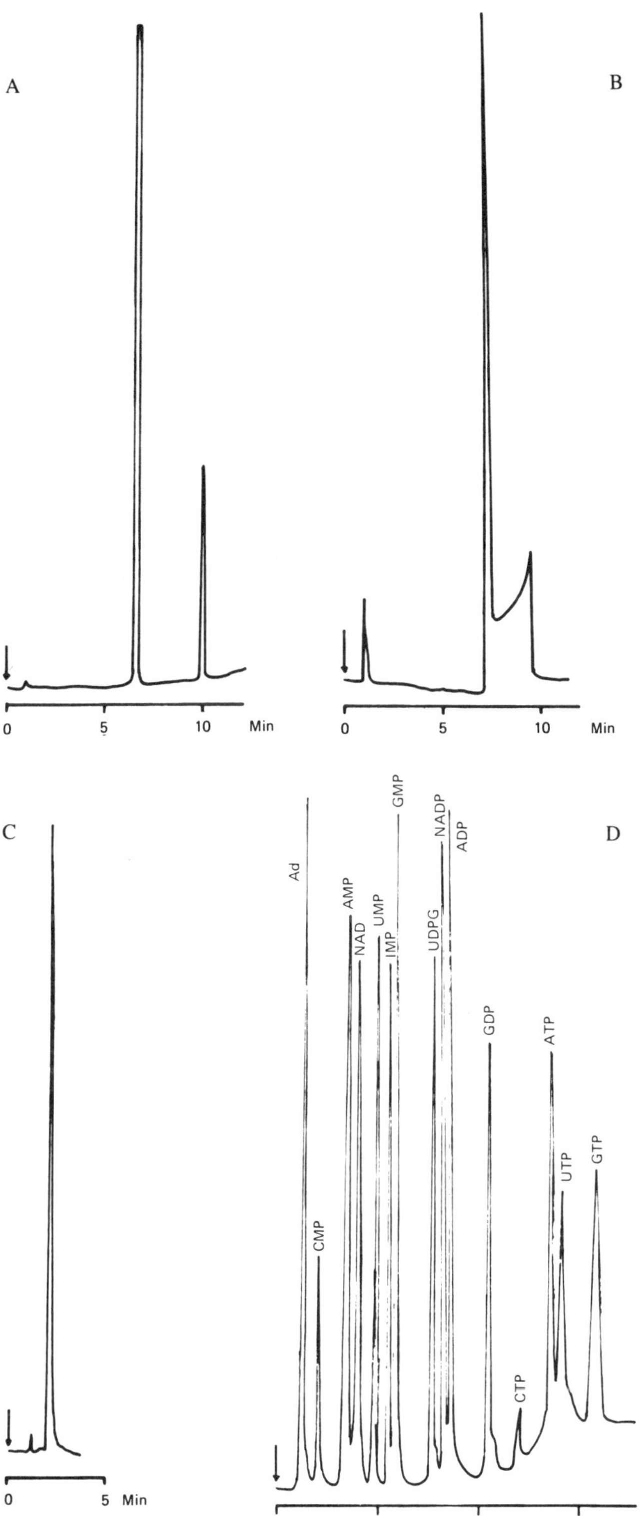

FIGURE 3. Pattern of interfering peaks caused by (A) Ficoll-Hypaque (4 μg injected) rt. 6.72 and 9.9 min; (B) EDTA (10 nmol injected) rt. 7.18 and 9.3 min; (C) ascorbic acid (25 nmol injected) rt. 1.84 min; (D) standard separation 1 nmol each nucleotide injected in the anion exchange gradient HPLC of nucleotides as per Perrett.[34] Details as in Figure 2.

calculated by comparing the $A_{650\ nm}$ reading before and after filtration. Finally, the formic solution was lyophilized prior to chromatography. Unfortunately, these authors[26] did not quote either ATP/ADP ratios or total nucleotide concentration in order to justify their conclusions.

In contrast to workers with animal tissues researchers in this field have traditionally, but not exclusively, used formic or acetic acids to release intracellular metabolites. In an early study[27] employing TLC separation of nucleotides, it was shown that NTPs were only partially released from *Escherichia coli* by a 6-min acetic acid extraction unless the cells were also freeze-thawed in the acid when recoveries were doubled. Lundin and Thore[28] compared ten methods for extracting nucleotides from a variety of bacteria, including *E. coli,* prior to a firefly assay. Methods tested included various acid and/or solvent extraction procedures. The results were compared using total nucleotide concentrations and energy charge (EC). The lowest EC was obtained with formic acid and chloroform, while TCA gave the highest values. Their findings, though, do not agree with the recent study of Olempska-Beer and Freese[29] who compared TCA and PCA plus Freon®/amine extractants with a procedure using formic acid and butanol. They harvested yeast cells by filtration and then extracted aliquots by the given procedures. Formic acid (1 mol/ℓ saturated with 1-butanol gave total nucleotide concentrations 50% higher than with TCA, but ATP/ADP ratios were 50% lower. This conflicting data may be due to degradation of NTPs during lyophilization of the formic acid extracts.

Similar procedures should be employed to harvest tissue culture cells and rapid filtration is possibly the best isolation technique. However, this author is unaware of a detailed investigation and workers have tended to use the more familiar mammalian procedures. Pogolotti and Santi[30] published a detailed study of the quantification of nucleotides in murine leukemia cells. They compared extractants and showed that ATP/ADP ratios were consistantly lower with PCA/KOH than with TCA. Maybaum et al.[31] harvested liver and lymphoma cells by centrifugation followed by homogenization in PCA, but they quote no values from which to calculate the energy charge. Chicken fibroblast nucleotides have also been extracted by a similar method.[32]

QUANTITATIVE ANALYSIS

Internal Standards

With complex extraction procedures such as those developed for nucleotides it is preferable to use an internal standard in order to correct for both recovery and analytical variability. The ideal internal standard should be chemically similar to the compounds of interest and be chemically stable. It should have similar detection characteristics as the unknown, and it should elute in a blank portion of the chromatogram, preferably near the middle (or near specific peaks of interest), and be well resolved from adjacent peaks. In order that it can accurately reflect the recovery of the unknowns, it must be added as near as possible to the start of the extraction procedure. This last point is difficult, if not impossible, to achieve for intracellular compounds such as nucleotides. It would be possible, at least with isolated cells, to load the cells with a radiolabeled nucleotide precursor and then quantify the recovery of the radiolabeled nucleotide. More often, a known quantity of a standard nucleotide that does not occur in the cells of interest is added along with the acid precipitant. The internal standard will now correct mainly for the dilution and sampling errors. The amount of internal standard added should be such that allowing for dilution, it will give a prominent well-resolved peak with a height of about 80% full scale under the normal analytical conditions employed. Various nucleotides such as CMP,[33] XMP, XDP,[34] and 5′-adenosine [βγ-methylene]triphosphate, which elutes between GDP and ATP on anion exchange,[35] have been employed as internal standards. CMP can be difficult to resolve from a number of frontally eluting compounds when an ion-exchange system is used. Nucleoside di- and triphosphates can degrade on storage and during extraction and are, therefore, not usually suitable. In the author's laboratory cyclic nucleotides, e.g., cXMP, although relatively expensive, have been found to be in many ways ideal. They have relatively good chemical stability, occur only in trace amounts in most tissues, and they elute in the center of many chromatographic systems.

When no suitable internal standard is available, a radiolabeled marker can be employed. We have used ^{14}C-guanine or inosine in some studies,[10] since with ion-exchange separations the bases are frontally eluted and it is an easy matter to collect the frontal peak in order to determine its radioactive content and calculate the recovery.[36]

Validation of Extraction Techniques

The quality of the extraction is assessed by calculating both the total amount of nucleotides extracted and the degradation of the nucleotides. Factors affecting the overall recovery of nucleotides from cells have already been discussed above. Changes in the in vivo concentration of nucleotides during extraction are usually indicated by changes in the ATP/ADP ratio or the energy of the cell. Atkinson[37] has defined the energy charge (EC) as follows:

$$EC = \frac{ATP + 1/2\ ADP}{ATP + ADP + AMP} \tag{1}$$

EC can be considered to be a quantitative estimate of the energy status of the cell. Its value should approach unity in a cell with maximum energy status and fall to zero in a fully depleted cell. In well-prepared tissue extracts the EC is generally in the range 0.85 to 0.95. For both these values it must be remembered that they are gross calculations, since the measurements employed are levels of total intracellular nucleotides and do not reflect any compartmentation which is almost certainly present in tissues.

Both EC and ATP/ADP ratio are conveniently determined using HPLC, since all the necessary elements are measured simultaneously and errors are minimized. Dean et al.[38] pointed out that for red cells with a "true" intracellular ATP/ADP ratio of 14, a 2% breakdown of ATP to ADP could cause a change in the ATP/ADP ratio of 24%, and the ADP level, provided there is little further breakdown to AMP, increases by 28%. Since the analytical errors were very small they concluded that the ATP/ADP ratio was an extremely sensitive measure of both the energy status of the cell and the rate of enzymatic and chemical degradation of the intracellular nucleotides. Changes in the ATP/AMP ratio for the red cell nucleotides of a number of species are shown in Figure 4. They also observed that changes in the ATP/ADP ratio could be paralleled by changes in other triphosphate/diphosphate pairs,[38] a finding later observed by Pogolotti and Santi.[30]

There are no definitive values for either the EC or the ATP/ADP ratios within cells, since all chemical methods of analysis require disturbing the equilibrium of the system and, thereby, changing the values. ^{31}P-NMR does not disrupt the cells since it can be performed on the whole tissue, but it measures only free nucleotides. A large percentage of the nucleotide pool as measured by TCA extraction appears to be protein bound, and therefore, not detected using NMR.

Ischemic changes in the ATP/ADP ratio can be rapid and both tissue and species dependent. Pogolotti and Santi[30] performed a detailed analysis on the ATP content and ATP/ADP ratio that they obtained on extracting L1210 cells. They were able to demonstrate a good correlation between the ATP/ADP ratio and the total ATP content of their cells, and suggested that the following relationship held true if there was minimal conversion of ADP to AMP:

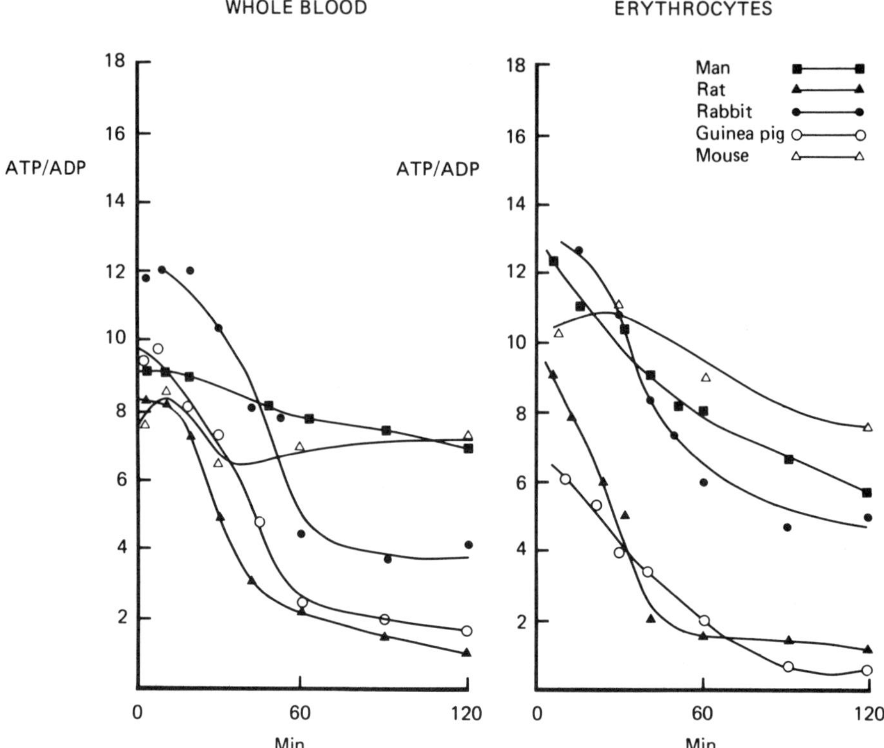

FIGURE 4. The effect of in vitro storage at room temperature on the ATP/ADP ratio in fresh whole blood (A) and separated erythrocytes (B) for different species. Values are the mean of two individual determinations per species.

$$\text{Fatp} = \frac{\text{Rm} + 1}{\text{Rm}} \times \frac{\text{Ro}}{\text{Ro} + 1} \tag{2}$$

where Fatp is the fraction of ATP/AMP + ADP + ATP, Rm is the true ATP/ADP ratio, and Ro is the found ATP/ADP ratio. For their cell line they observed a maximum Ro of 22 and a Fatp of 0.96 and that the Fatp value was almost independent of the ATP/ADP ratio over the range 8 to 25. They concluded that for their system ratios below 6 could be considered suspect.

Such values should not be used for all tissues and the cutoff ratios must be determined for individual tissues and preparations. It is not uncommon to find that with practice at isolating and extracting a tissue the ATP/ADP ratios increase dramatically. The first extractions of some organs such as liver often give values below 1, whereas an experienced worker can consistantly achieve ratios as high as 7. Such variability is also, unfortunately, reflected in the published literature (see comparative table in Reference 18 and the tables in this chapter). In all cases the biochemistry of the tissue must be kept in mind. The ATP/ADP of platelets is approximately 1, since they contain a large stored pool of ADP which is not normally in equilibrium with ATP.

Although EC should give a more comprehensive measure of the energy status in an extract, it has been used by fewer workers since it is more difficult to determine, particularly when minor changes are being measured and AMP is not accumulating. EC has been employed by Ericson and De Verdier[39] to evaluate specimen handling techniques for red cells, Hultman et al.[21] in their study on human liver, and Lundin and Thore[28] in their study on bacteria.

Stability of Nucleotide Extracts

Nucleotides are readily hydrolyzed in acidic solutions, but are relatively stable over the pH 5 to 8. It is for this reason that PCA and TCA extracts are processed to completely remove the excess acid so that they can be stored at neutral pH. Brown and Miech[6] showed that storage of unneutralized TCA extracts at −20°C resulted in a 36 and a 64% reduction of ATP after 1 week and 1 month storage, respectively, whereas a neutralized solution was totally stable even after 4 months at −20°C. Experience in this laboratory would suggest that storage at this temperature of neutral extracts for up to 6 months is possible without seriously changing either the quantitative or qualitative appearance of the chromatogram. Some nucleotides such as NADP are even more labile, and various degradation products can be formed during extraction and storage that may interfere with the accurate quantification of other nucleotides.[33] Provided the samples are neutralized, nucleotide extracts can be kept at room temperature during the analyses. This is particularly important when auto-injectors are being used and samples may wait for many hours before being injected. Longer-term storage at room temperature shows changes that are probably due to bacterial contamination. In this laboratory we have observed that the first changes occur to the frontal peak, i.e., the nucleosides and bases, which decrease significantly after about 1 week at room temperature.

Sources of Error in Nucleotide Determinations

Without doubt the major source of error in nucleotide determinations is incorrect sample preparation leading to degradation of nucleotides due to warm ischemia. The avoidance of such changes and the identification of those changes have already been discussed in detail. Other factors may also lead to erroneous results, some of which are outlined in this section.

Chemical Impurities

The analytical sensitivity and resolution of HPLC can often detect impurities and/or breakdown products in standard materials brought from reputable supply companies, since the analytical techniques used in the laboratory are sometimes better than those employed in the company's quality control laboratory. This is particularly so with such readily hydrolyzed compounds as nucleotide triphosphates. Brenton et al.[40] reported that commercial grades of some nucleotides were impure. In this laboratory we commonly observe that adenine and guanine nucleotide standards contain trace amounts of their degradation products, but with pyrimidine nucleotides these contaminants can in some materials exceed 10% of the total material. Riss et al.[15] reported the following percentages for nucleoside diphosphate in nucleoside triphosphate standards: 2.4, 1.7, 4.9, and 4.6% in ATP, CTP, GTP, and UTP, respectively, and found that these levels doubled during acid extraction. Yoshika et al.[41] found that commercial samples of ATP contained between 0.1 and 0.9% dATP and the same group found that the reverse also applied.[42] They described methods to purify the commercial samples.

UV-absorbing impurities may also arise from other chemicals in either the eluent or the extraction procedure. The problems associated with the phosphate gradient used to elute anion exchange columns have been known for some time.[43] The quality of the water employed is also important.[44] Fortunately, it is now possible to find suitable buffer salts among the highest grade materials sold by some manufacturers.[44] Other UV-absorbing salts have been reported. Both Ficoll/Hypaque and EDTA may contaminate blood cells and examples of their chromatographic products are shown in Figures 3A and 3B, respectively. Some non-nucleotides are also observed, for example, ascorbic acid elutes close to CMP on the ion-exchange system (Figure 3C) and can even be quantified in lymphocytes.[45]

Sampling Errors

For meaningful estimates of intracellular nucleotides, the sample must be homogeneous.

The percentage contamination of blood cell preparations can be easily determined, but it is almost impossible to freeze-clamp a uniform area of some tissues. The best approach is to use tongs with the smallest heads compatible with the tissue and the analytical system. Even so, we have been unable to separate gut villi from the mucosa without severely reducing the ATP/ADP ratio. With blood cells it is possible to devise correction factors for contamination. Tissues such as liver and kidney, when freeze-clamped, can contain substantial amounts of blood. Roberts and Bessman[23] used mathematical procedures to correct for the trapped erythrocytes in their study of fetal heart tissue. They assumed that 2,3-DPG was marker for the numbers of red cells, and by relating its concentration to that of isolated red cells they calculated that the nucleotide concentration in the heart tissue was approximately 25 to 35% higher than suggested by the initial measurements. Determination of hemoglobin or some other red cell marker could also be used to make this correction. Platelet preparations usually contain some red cells, and Goday et al.[46] have shown that literature data suggesting the accumulation of dATP in lymphocytes from patients with adenosine deaminase deficiency are incorrect. The apparent dATP is due to the presence of platelets which do accumulate dATP in the lymphocyte preparations. This degree of contamination also lowered the ATP/ADP ratio from a true value of 17 to a measured value of 2.3. In all cases, though, one must balance the advantages of preparing a more homogeneous sample against the nucleotide degradation which may occur in the time taken to achieve a significant improvement in the sample.

Few studies have addressed another aspect of heterogeneity which is caused by changes in nucleotide composition due to differences in the age of tissues and cells, although it has long been known that recticulocytes contain more ATP than the mature erthrocyte. Hjelm and Arturson[18] found that the concentration of adenine nucleotides in the liver and skeletal muscle of Wistar rats increased with body weight (i.e., age), whereas the concentration in erthrocytes decreased. The small changes in the water content of the tissue (e.g., 5% decrease in muscle) were insufficient to explain the increases in the level of ATP in muscle (approximately 50%) or liver (approximately 30%).

Both quantitative and qualitative differences seen in the nucleotide content of tissues from different species have been documented. For example, Bartlett[47] has comprehensively surveyed the nucleotide content of red cells from a wide variety of animals. Although such differences are to be expected, Dean et al.[38] found marked differences in the rate of degradation of red cell nucleotides between species. The differences were such that extraction procedures developed for the human erythrocyte were totally inadequate for some animals such as the rat, where the ATP/ADP ratio could decrease by 80% in 45 min (Figure 4). This type of difference may well apply to other tissues and species, and it will be necessary to evaluate extraction techniques independently.

The differences in nucleotide levels caused by either physiological processes or disease states are beyond the scope of the present work. Some changes were outlined in an early review by Mandel,[4] and this author is unaware of any more recent comprehensive discussion which includes the major advances due to HPLC in the last 10 years.

Quantitative Errors

These may arise at any of the sampling and/or manipulative stages of the extraction. For example, it is possible, but inaccurate, to allow for intracellular water or water trapped around cells by calculation alone. The use of internal standard will adequately account for such errors, as well as correct for chromatographic errors. An appropriate internal standard may also correct for losses onto precipitates such as $KClO_4$, losses into the ether phase with TCA extraction, or losses into the Freon®/amine mixture.

The final calculated nucleotide concentrations for tissues should be expressed in a variety of forms such as per gram wet weight, per gram dry weight, per milligram protein, etc., and the method giving the lowest standard deviation employed. Similarly, for blood cells use per milliliter packed cells, per 10^6, or per milligram protein. In all cases, suitable samples for the calculation of the appropriate fractions should be collected.

ISOLATION OF MINOR NUCLEOTIDES

Deoxynucleotides

dNXPs can be isolated from cells by the same procedures and with the same precautions as the ribonucleotides. dNXPs are less acid stable than ribonucleotides, but no information on their relative biochemical stability is available. The major problems are analytical since, although they elute close to their respective ribonucleotide and can be readily resolved by a variety of systems, they are usually present in very low concentrations. To overcome this difference in relative concentration, methods involving the selective degradation of ribonucleotides prior to HPLC have been devised.

Ritter and Bruce[48] adapted for HPLC the TLC method of Yeagin,[49] which employed periodate to react selectively with the 2′3′-*cis*-diol of ribonucleotides. The reaction products were then readily resolved from the unreacted dNTPs following the removal of excess periodate with rhamnose. Pestell and Woodland[50] used an additional step involving the addition of methylamine to cleave the *N*-glycosidic bond, thereby forming the free base. It was shown by Garrett and Santi[51] that the resultant mixture could be injected directly onto an anion-exchange column for quantification of the nucleotides. Martinez-Valdez and colleagues[52] found that the direct injection of this mixture altered the elution times of the deoxyribose compounds. Tanaka et al.[42] found that this procedure led to the formation of an impurity which interfered with the anlysis of dTTP and dATP, and that up to 20% of the dGTP could be destroyed. The first problem was overcome by modifying the order of addition of the reagents, while the addition of excess deoxyguanosine to the reaction mixture gave 100% recovery of dGTP. In outline their final procedure is to add 20 μℓ 20 mmol/ℓ deoxyguanosine and 20 μℓ 0.2 mol/ℓ $NaIO_4$ to 80 μℓ of cell extract. The mixture is vortexed and centrifuged for 20 sec followed by incubation at 37°C for 2 min. Then, 2 μℓ of 1 mol/ℓ rhamanose are added followed by 30 μℓ 4 mol/ℓ methylamine (neutralized to pH 6.5 with phosphoric acid). Mix, centrifuge for 20 sec, and incubate at 37°C for 30 min; then cool on ice. The sensitivity of such a procedure is about 30 pmol injected.

Other procedures for selectively measuring the deoxynucleotides involve anion-exchange fractionation of nucleotide pairs, followed by conversion to nucleoside pairs using acid phosphatase. The deoxynucleoside and the ribonucleoside of both purine[53] and pyrimidines[54] are then separated by reversed-phase HPLC.

Cyclic Nucleotides

Although HPLC is well suited to the separation of cyclic nucleotide mixtures, in practice it is not applicable to crude acid extracts of tissues. This is due to the overwhelming concentrations of noncyclic nucleotides, which prevent the identification of the cyclic nucleotide peaks. The cyclic nucleotides are extracted from tissues etc. by procedures similar to those previously described. In order to measure cAMP, Brooker[55] adsorbed the ribonucleotides onto the precipitate of barium sulfate formed by adding $Ba(OH)_2$ and $ZnSO_4$ to the sample. In that way it is possible to detect as little as 1 pmol of cAMP using current equipment and microparticulate column packings. It is possible to assay cAMP in raw extracts by HPLC combined with fluorometric[56] and, possibly, cGMP using electrochemical detection.[57]

Polynucleotides

A number of unusual highly phosphorylated nucleotides have been described during the last 20 years. Recently, various authors have included an HPLC step in either direct quantitative methods or in qualitative separations. In general, the extraction procedures already described have been employed prior to either anion-exchange[35,58] or reversed-phase chromatography.[59] Lagosky and Chang[60] found that the most common method, i.e., 1 mol/ℓ formic acid at 0°C, to release guanosine 5′-diphosphate,3′-diphosphate (ppGpp) from *E. coli*, resulted in the breakdown of approximately 50% of the polynucleotide to ppGp and GDP. They compared a variety of similar extractants and showed that the degradation can be largely prevented by a freeze/thawing at neutral pH in the presence of lysozyme. Whether the same degree of degradation affects other polynucleotides, such as diadenosine triphosphate (Ap3A) or diadenosine tetraphosphate (Ap4A) which occurs in human platelets, has not been reported.

PRELIMINARY FRACTIONATION PRIOR TO HPLC

Because of the relatively low sensitivity of the UV detection systems most frequently employed to measure nucleotides, it is sometimes necessary to concentrate the NXPs before injection onto the analytical column. Freeze drying is commonly employed, but the resulting solution is often high in salts which disturb the analytical separation. There is also a possibility that losses of nucleotides may occur. NTPs and NDPs can be concentrated from dilute solutions by coprecipitation with calcium fluoride. Following filtration, the nucleotides are released from the precipitate with 0.5 mol/ℓ sulfuric acid.[61]

Because of the complexity of acid extracts which contain not only nucleotides, nucleosides, and bases, but also other UV-absorbing compounds such as tryptophan and ascorbic acid, some workers have preferred to fractionate these groups of compounds before the HPLC analysis. Approaches to separating these groups include ligand exchange, gel filtration, boronate columns, and ion-exclusion columns.

Ligand exchange chromatography on Cu^{2+} loaded Chelex® 100 columns,[62] which adsorb nucleosides and bases but frontally elute nucleotides, was used by Brown et al.[33] in their studies. They claimed that such fractionation improved the life of the analytical column, but like all these procedures, it involved freeze drying large (approximately 14 mℓ) fractions of ligand column eluate (see Figure 5).

Boronate affinity chromatography has also been used to enhance the value of the analytical HPLC separating crude extracts.[26] At alkaline pH, boronate groups complex with the *cis*-diols of ribonucleosides and ribonucleotides, while dNXPs and cyclic nucleotides are not retained.[63] Lothrop and Uziel[64] adsorbed acid extracts of nucleotides onto 1-mℓ silica (Sep-pak®) cartridges. Bases and nucleosides are not adsorbed from 90:10 acetonitrile/water mixtures, but nucleotides are retained and they can be, subsequently, eluted with water.

TYPICAL INTRACELLULAR NUCLEOTIDE LEVELS

A comprehensive listing of all published nucleotide data would not be possible, since many studies have only determined a limited group of nucleotides such as the adenosine compounds, while many other studies have employed unsatisfactory procedures. For examples of earlier determinations the reader should consult the review by Mandel.[4]

Tables 3 to 7 attempt to draw together some of the more recently obtained values for tissue nucleotides, particularly those which have been determined by HPLC, although some good examples which employed other techniques have also been included for comparative purposes. The tables indicate the methods the authors employed. If the methods are chromatographic the total number of peaks measured or reported is given. The majority of the chromatographic assays used ion-exchange HPLC. This undoubtedly reflects the inability of early reversed-phase methods to adequately resolve the nucleotides in complex physiological samples.

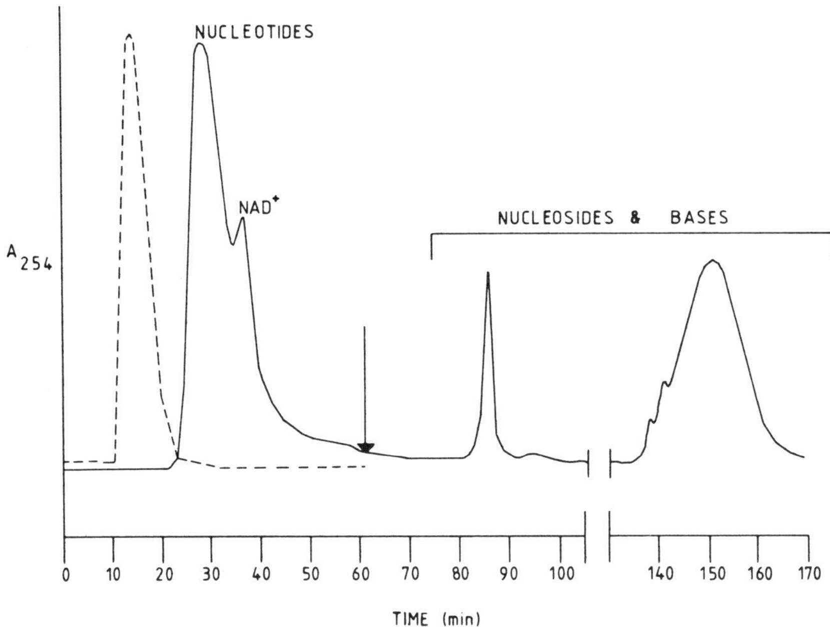

FIGURE 5. Preliminary fractionation of extract of rat liver (100 mg) on Chelex® 100. 20 × 9 mm of Cu^{2+}-loaded Chelex® 100 was eluted with water 0.7 mℓ/min. At 62 min the eluent was changed to 2.5 mol/ℓ aq. ammonia. The broken line shows the elution with water at 1.4 mℓ/min. (From Brown, E. G., Newton, R. P., and Shaw, N. M., *Anal. Biochem.*, 123, 378, 1982. With permission.)

A number of studies of species differences have been published. Bartlett has surveyed the red cell nucleotide patterns from many species in a series of publications[47,65-70] (using traditional ion-exchange chromatography combined with the measurement of phosphates using a molybdate reaction[71]). Applying similar separation techniques Seki and Saito[72] quantified nucleotides in hepatic tissues of eight aquatic animals. Many of these published studies are characterized by low ATP/ADP ratios which must cast some doubt on the validity of the results. Using enzymatic assays, Beis and Newsholme[20] measured the adenine nucleotides and some other metabolic intermediates in the resting muscles from 23 different species of vertebrates and invertebrates. Although HPLC has the potential to rapidly qualitatively and quantitatively analyze many samples and would be ideal for such species comparison, few studies have been reported. Using an early nucleotide analyzer with pellicular ion exchanges, Brown's group[73] determined the nucleotide content of whole blood in 16 species.

Accurate comparative data on individual tissues in the same species are also difficult to find. Scholar et al.[74] measured the nucleotide content of the formed elements of human blood. Other studies are quoted in the tables.

Table 3
NUCLEOTIDE CONTENT OF VARIOUS WHOLE TISSUES

Tissue	AMP	ADP	ATP	GMP	GDP	GTP	UTP	CTP	Total	ATP/ADP	Units	Method	Ref.
Liver													
Human	880 ± 600	3800 ± 600	8230 ± 1960	—	—	—	—	—	3	9.35	nmol/g dry wt	Enzyme	21
	500 ± 200	980 ± 200	1910 ± 560	—	—	—	—	—	3	1.9	nmol/g wet wt	Enzyme	75
Rat	221 ± 58	852 ± 53	2719 ± 143	21 ± 3	56 ± 14	292 ± 15	249 ± 28	27 ± 4	17	3.3	nmol/g wet wt	RP-LC	76
	302 ± 30	958 ± 60	1880 ± 140	—	—	—	—	—	3	1.95	nmol/g wet wt	Enzyme	77
		796 ± 70	3220 ± 154	—	87 ± 10	391 ± 19	—	—	4	4.2	nmol/g wet wt	Enzyme	78
	260	1340	2740	—	—	—	—	—	3[a]	2.0	nmol/g wet wt	Enzyme	20
Mouse	320 ± 40	600 ± 50	2780 ± 13	—	—	—	—	—	3	4.6	nmol/g wet wt	Lucifr.	79
Kidney													
Rat	250	960	1710	—	—	—	—	—	3[a]	2.0	nmol/g wet wt	Enzyme	20
Heart													
Rat	2.3 ± 0.2	6.3 ± 0.3	20.3 ± 0.7	—	0.54 ± 0.1	2.49 ± 0.2	—	1.3 ± 0.2	10	3.2	μmol/g prot	HPLC-IE	23
	90	850	4480	—	—	—	—	—	3[a]	8.6	nmol/g wet wt	Enzyme	20
Lung													
Mouse	280 ± 30	600 ± 40	1650 ± 7	—	—	—	—	—	3	4.6	nmol/g wet wt	Lucifr.	79
Brain													
Rat	80 ± 20	960 ± 99	8830 ± 13010	—	340 ± 90	1860 ± 440	640 ± 99	—	14	10.3	nmol/g dry wt	HPLC-IE	24
Mouse	290 ± 36	800 ± 41	2110 ± 58	—	—	—	—	—	3	2.6	nmol/g wet wt	Lucifr.	79
Muscle (resting)													
Human	(0.1—0.4)	(2.0—4.2)	(18.5—28.2)	—	—	—	—	—	3[b]		μmol/g dry wt	Enzyme	80
Rat	50 ± 10	720 ± 40	6230 ± 340	—	—	—	—	—	3[a]	8.6	nmol/g wet wt	Enzyme	20
Mouse	100	900	4800	—	—	—	—	—	3	5.4	nmol/g wet wt	Enzyme	81
	30 ± 10	550 ± 50	4990 ± 320	—	—	—	—	—	3[a]	9.4	nmol/g wet wt	Enzyme	20
	38 ± 7	732 ± 125	6982 ± 1194	—	—	—	—	—	10	9.6	nmol/g wet wt	HPLC-IE	10

Intestine (rat)												
Jejunum	382 ± 102	3314 ± 476	8999 ± 1913	245 ± 133	923 ± 89	1744 ± 340	—	—	22	nmol/g dry wt	HPLC-IE	82
Ileum	482 ± 238	3124 ± 682	9193 ± 1660	139 ± 63	797 ± 99	1586 ± 324	—	—	22	nmol/g dry wt	HPLC-IE	82

Note: Total refers to number of nucleotides that could be measured by the published assay or, in the case of HPLC methods, the number of nucleotides identified in the calibration solution. Under "methods", enzyme refers to a variety of enzymatic methods (see paper for details); Lucifr. is the standard luciferase methods; HPLC-IE nearly always refers to anion-exchange gradient elution procedures; RP-LC = reversed-phase LC. Means and SD are normally, in some cases, quoted ranges, and in others just single values. A blank implies not given in the reference and presumed not measured.

[a] Includes glycolytic intermediates.

[b] Normal range given in brackets.

Table 4
NUCLEOTIDES IN THE ELEMENTS OF BLOOD

Cell	AMP	ADP	ATP	GMP	GDP	GTP	UTP	CTP	Total	ATP/ADP	Units	Method	Ref.
Erythrocytes													
Man	50 ± 10	270 ± 80	1740 ± 163	—	20 ± 10	10 ± 1	—	—	7	6.4	nmol/mℓ RBC	HPLC-IE	74
	10 ± 3	114 ± 24	1278 ± 127	—	25 ± 5	131 ± 20	—	—	10	11.2	nmol/mℓ RBC	HPLC-IE	36
Rat	—	400	1900						2	4.7	nmol/mℓ RBC	Enzyme	18
	250	530	1130	—	—	0	0	0	22	2.15	nmol/mℓ RBC	Enzyme	83
	15	87	651	—	21	108	—	—	10	7.7	nmol/mℓ RBC	Enzyme	36
Rabbit	21	90	1036	—	20	230	—	—	10	11.2	nmol/mℓ RBC	Enzyme	36
Guinea pig	9	125	752	9	44	2	0	—	22	6.0	nmol/mℓ RBC	HPLC-IE	Author
Mouse	17	163	1133	1.4	17	231	67	0	22	9.9	nmol/mℓ RBC	HPLC-IE	Author
Pigeon	18	79	2345	21	14	296	195	0	22	21.2	nmol/mℓ RBC	HPLC-IE	
Leurocytes													
Man	70 ± 20	140 ± 130	990 ± 173	—	36 ± 5	300 ± 70	68 ± 10		7	7.1	nmol/10^9cell	HPLC-IE	74
Lymphocytes													
Man	30	170	2880	—	60	460	370	80	10	17.4	nmol/10^9	HPLC-IE	45
	(10—70)[a]	(90—270)[a]	(1440—4450)[a]		(40—120)[a]	(110—870)[a]	(170—450)[a]						
	43 ± 26	302 ± 64	855 ± 15	9 ± 3	—	—	—	—	6	2.8	nmol/10^9	HPLC-IE	84
	25 ± 11	53 ± 40	500 ± 140	7 ± 5	22 ± 10	128 ± 47	102 ± 31	73 ± 44	21	13.7	nmol/10^9	HPLC-IE	85
Granulocytes													
Man	—	90 ± 10	750 ± 130	—	12 ± 6	160 ± 30	27 ± 7		7	8.3	nmol/10^9 cell	HPLC-IE	74
Platelets													
Man	7.7	59	96	1.8	9.9	17.7	7.9	1.99	10	1.6	nmol/10^9	HPLC-IE	45
	(2.7—12.3)[a]	(35—77)[a]	(56—137)[a]	(0.7—4.6)[a]	(4—17)[a]	(8—33)[a]	(4—14)[a]	(1—3.8)[a]					
	4.9	19	32	2.7	4.6	4.9	2.0	1.4	14	1.6	nmol/10^9	HPLC-IE	86
	—	35 ± 8	57 ± 7	—	9 ± 1	9 ± 1	68 ± 10		7	1.7	nmol/10^9	HPLC-IE	74
	3.2 ± 1.4	25 ± 7	37.8 ± 7	1.5 ± 0.6	3.8 ± 0.7	4.5 ± 0.7	0.7 ± 0.3	3 ± 0.1	12	1.5	nmol/10^9	HPLC-IE	87
Plasma													
Man	2.6	3.4	1.5	—	—	—	—	—	10	0.4	nmol/mℓ	RP-LC	24

Note: Total refers to number of nucleotides that could be measured by the published assay or, in the case of HPLC methods, the number of nucleotides identified in the calibration solution. Under "methods", enzyme refers to a variety of enzymatic methods (see paper for details); HPLC-IE nearly always refers to anion-exchange gradient elution procedures; RP-LC = reversed-phase LC. Means and SD are normally, in some cases, quoted ranges, and in others just single values. A blank implies not given in the reference and presumed not measured.

[a] Normal range given in paper.

Table 5
NUCLEOTIDE CONTENT OF ISOLATED CELLS IN CULTURE

Cell	AMP	ADP	ATP	GMP	GDP	GTP	UTP	CTP	Total	ATP/ADP	Units	Method	Ref.
Liver													
Rat	490 ± 13	742 ± 69	8,740 ± 225	274 ± 7	620 ± 20	2,075 ± 128	557 ± 138	935 ± 209	12	11.8	nmol/10^9 cell	HPLC-IE	15
	36	264	447		70	57	49	79	17	1.7	nmol/10^9 cell	HPLC-IE	63
	860	1,542	3,350						5	2.2	nmol/g prot	HPLC-IE	88
Fibroblasts													
Man	377 ± 93	1,678 ± 288	9,783 ± 869	354 ± 107	—	—	—	—	9	5.9	nmol/10^9 cell	RP-LC	89
	100	1,090	14,150	10	180	2,060	2,840	220	37[a]	13	nmol/10^9 cell	HPLC-IE	90
Chicken	—	580	2,680	—	—	250	150	61	13	4.6	nmol/10^9 cell	TLC-UV	31
Lymphona													
Mouse	S49 36	124	1,300	1.8	23	227	560	166	20[b]	10.5	nmol/10^9 cell	HPLC-IE	52, 53
	—	—	2,810	—	—	484	998	465	8[b]	—	nmol/10^9 cell	HPLC-IE	50
Lymphoblasts													
W1-L2	—	570 ± 40	4,230 ± 230	50 ± 10	170 ± 20	1,260 ± 80	1,620 ± 170	530 ± 80	12	7.4	nmol/10^9 cell	HPLC-IE	39

Note: Total refers to number of nucleotides that could be measured by the published assay or, in the case of HPLC methods, the number of nucleotides identified in the calibration solution. Under "methods", enzyme refers to a variety of enzymatic methods (see paper for details); HPLC-IE nearly always refers to anion-exchange gradient elution procedures; RP-LC = reversed-phase LC. Means and SD are normally, in some cases quoted ranges, and in others just single values. A blank implies not given in the reference and presumed not measured.

[a] Includes bases and nucleosides.

[b] Includes deoxynucleotides by periodate method.

Table 6
NUCLEOTIDE CONTENT OF SOME MISCELLEANEOUS CELLS ETC.

Cell	AMP	ADP	ATP	GMP	GDP	GTP	UTP	CTP	Total	ATP/ADP	Units	Method	Ref.
Growth plate cartilages													
Chicken	190	1490	3370	150	250	840	—	—	7	2.3	nmol/g dry wt	HPLC-IE	91
Wound-healing tissue													
Rat	—	3500	5500	—	—	—	—	—	7	1.6	nmol/g prot	HPLC-IE	37
Ova													
Mouse	0.1	0.56	1.52	—	—	—	—	—	3	2.7	pmol per ova	Enzyme	92
Chloroplasts													
Spinach	—	780 ± 210	2630 ± 310	—	—	—	—	—	3	0.3	nmol/g chlorophyll	HPLC-IE	93
Seeds													
Wheat[a]	137 ± 10	106 ± 5	1092 ± 32	—	35 ± 3	246 ± 13	194 ± 6	168 ± 5	16	10	pmol per embyro	HPLC-IE	94

Note: Total refers to number of nucleotides that could be measured by the published assay or, in the case of HPLC methods, the number of nucleotides identified in the calibration solution. Under "methods", enzyme refers to a variety of enzymatic methods (see paper for details); HPLC-IE nearly always refers to anion-exchange gradient elution procedures; Means and SD are normally, in some cases, quoted ranges, and in others just single values. A blank implies not given in the reference and presumed not measured.

[a] After 5.5 hr germination at room temp.

Table 7
NUCLEOTIDE CONTENT OF BACTERIA

Type	AMP	ADP	ATP	GMP	GDP	GTP	UTP	CTP	Total	$\frac{ATP}{ADP}$	Units	Method	Ref.
E. coli	—	—	6500	—	—	3600	3700	1400	4	—	nmol/g cell	TLC-UV	26
Yeast (*S. cerevisiae*)	—	351 ± 10	953 ± 25	—	108 ± 5	233 ± 14	289 ± 16	207 ± 6	8	2.7	pmol/OD_{600}	HPLC-IE	28

Note: Total refers to number of nucleotides that could be measured by the published assay or, in the case of HPLC methods, the number of nucleotides identified in the calibration solution. Under "methods", enzyme refers to a variety of enzymatic methods (see paper for details); TLC-UV = thin layer chromatography with UV detection. HPLC-IE nearly always refers to anion-exchange gradient elution procedures; RP-LC = reversed-phase LC. Means and SD are normally, in some cases, quoted ranges, and in others just single values. A blank implies not given in the reference and presumed not measured.

REFERENCES

1. **Henderson, J. F. and Paterson, A. R. P.,** *Nucleotide Metabolism, An Introduction,* Academic Press, New York, 1973.
2. **Hutchinson, W. C. and Munro, H. N.,** The determination of nucleic acids in biological materials, *Analyst,* 86, 768, 1961.
3. **Munro, H. N. and Fleck, A.,** The determination of nucleic acids, *Methods Biochem. Anal.,* 14, 113, 1964.
4. **Mandel, P.,** Free nucleotides in animal tissues, *Prog. Nucl. Acid Res. Mol. Biol.,* 3, 299, 1964.
5. **Maickel, R. P.,** Separation science applied to analyses on biological samples, in *Drug Determination in Therapeutic and Forensic Contexts,* Vol. 14, Reid, E. and Wilson, I. D., Eds., Plenum Press, New York, 1984, 3.
6. **Brown, P. R. and Miech, R. P.,** Comparison of cell extraction procedures for use with high pressure liquid chromatography, *Anal. Chem.,* 44, 1072, 1972.
7. **Brown, P. R.,** Comparison of cell extraction procedures for use with high pressure liquid chromatography; addendum, *Anal. Chem.,* 47, 784, 1975.
8. **Hartwick, R. A., Van Haverbeke, D., McKeag, M., and Brown, P. R.,** Sample preparation techniques prior to HPLC analysis of serum nucleosides and their bases, *J. Liq. Chromatogr.,* 2, 725, 1979.
9. **Blanchard, J.,** Evaluation of the relative efficacy of various techniques for deproteinizing plasma samples prior to high performance liquid chromatographic analysis, *J. Chromatogr.,* 226, 455, 1981.
10. **Lush, C., Rahim, Z. H. A., Perrett, D., and Griffiths, J. R.,** A micro-procedure for extracting tissue nucleotides for analysis by HPLC, *Anal. Biochem.,* 93, 227, 1979.
11. **North, T. W., Bestwick, R. K., and Bennett, L. L.,** Detection of activities that interfere with the enzymatic assay of deoxyribonucleotide 5′-triphosphates, *J. Biol. Chem.,* 255, 6640, 1980.
12. **Khym, J. X.,** An analytical system for rapid separation of tissue nucleotides at low pressures on conventional anion exchangers, *Clin. Chem.,* 21, 1245, 1975.
13. **Chen, S.-C., Brown, P. R., and Rosie, D. M.,** Extraction procedures for use prior to HPLC nucleotide analysis using microparticle chemically bonded packings, *J. Chromatogr. Sci.,* 15, 218, 1977.
14. **Van Haverbeke, D. A. and Brown, P. R.,** Optimization of a procedure for extraction of nucleotides from plasma and erthrocytes prior to HPLC analysis, *J. Liq. Chromatogr.,* 1, 507, 1978.
15. **Riss, T. L., Zorich, N. L., Williams, M. D., and Richardson, A.,** A comparison of the efficiency of nucleotide extraction by several procedures and the analysis of nucleotides from extracts of liver and isolated hepatocytes by HPLC, *J. Liq. Chromatogr.,* 3, 133, 1980.
16. **Faupel, R. P., Seitz, H. J., Tarnowski, W., Thiemann, V., and Weiss, C. H.,** The problem of tissue sampling from experimental animals with respect of freezing technique, anoxia, stress and narcosis, *Arch. Biochem. Biophys.,* 148, 509, 1972.
17. **Wollenberger, A., Ristau, O., and Schoffa, G.,** Eine einfache technik der extrem schnellen abkuhlung großerer gewebestucke, *Pflugers Arch.,* 270, 399, 1960.
18. **Hjelm, M. and Arturson, G.,** The amount of adenine nucleotides and glycoytic intermediates in erthrocytes, liver and muscle tissue correlated with the body weight (age) in Wistar rats, *Upsala J. Med. Sci.,* 87, 99, 1982.
19. **Bucher, N. L. R. and Swaffield, M. N.,** Nucleotide pools and [6-^{14}C] orotic acid incorporation in early regenerating rat liver, *Biochem. Biophys. Acta,* 129, 445, 1966.
20. **Beis, I. and Newsholme, E. A.,** The contents of adenine nucleotides, phosphagens and some glycolytic intermediates in resting muscles from vertebrates and invertebrates, *Biochem. J.,* 152, 23, 1975.
21. **Hultman, E., Nilsson, L. H., and Sahlin, K.,** Adenine nucleotide content of human liver; normal values and fructose-induced depletion, *Scand. J. Clin. Lab. Invest.,* 35, 245, 1975.
22. **Fernando, A. R., Armstrong, D. M. G., Griffiths, J. R., Hendry, W. F., O'Donoghue, E. P. N., Perrett, D., Ward, J. P., and Wickham, J. E. A.,** Enhanced preservation of the ischaemic kidney with inosine, *Lancet,* March 13th, 555, 1976.
23. **Roberts. C. M. and Bessman, S. P.,** Measurement of phosphate esters in extracts of fetal rat heart by automated high pressure liquid chromatography, *Biochem. Biophys. Res. Commun.,* 93, 617, 1980.
24. **Shmukler, H. W.,** The rapid chromatographic analysis of the free nucleotides from rat brain, *J. Chromatogr. Sci.,* 10, 38, 1972.
25. **Harkness, R. A., Coade, S. B., and Webster, A. D. B.,** ATP, ADP and AMP in plasma from peripheral venous blood, *Clin. Chim. Acta,* 143, 91, 1984.
26. **Payne, S. M. and Ames, B. N.,** A procedure for rapid extraction and high pressure liquid chromatographic separation of the nucleotides and other small molecules from bacterial cells, *Anal. Biochem.,* 123, 151, 1982.
27. **Nazar, R. N., Lawford, H. G., and Wong, J. T.-F.,** An improved procedure for extraction and analysis of cellular nucleotides, *Anal. Biochem.,* 35, 305, 1970.

28. **Lundin, A. and Thore, A.,** Comparison of methods for extraction of bacterial adenine nucleotides determined by firefly assay, *Appl. Microbiol.*, 30, 713, 1975.
29. **Olempska-Beer, Z. and Freese, E. B.,** Optimal extraction conditions for high performance liquid chromatographic determination of nucleotides in yeast, *Anal. Biochem.*, 140, 236, 1984.
30. **Pogolotti, A. L. and Santi, D. V.,** High-pressure liquid chromatography ultraviolet analysis of intracellular nucleotides, *Anal. Biochem.*, 126, 335, 1982.
31. **Maybaum, J., Klein, F. R., and Sadee, W.,** Determination of pyrimidine ribotide and deoxyribotide pools in cultured cells and mouse liver by high performance liquid chromatography, *J. Chromatogr.*, 188, 149, 1980.
32. **Colby, C. and Edlin, G.,** Nucleotide pool levels in growing, inhibited and transformed chick fibroblast cells, *Biochemistry*, 9, 917, 1970.
33. **Brown, E. G., Newton, R. P., and Shaw, N. M.,** Analysis of the free nucleotide pools of mammalian tissues by high-pressure liquid chromatography, *Anal. Biochem.*, 123, 378, 1982.
34. **Pruneau, D., Wulfert, E., Pascal, M., and Baron, C.,** High performance liquid chromatographic procedure for measuring ATP and ADP levels in tissue microbiopsy: application to rat wound healing proliferative tissue, *Anal. Biochem.*, 119, 274, 1982.
35. **Flodgaard, H. and Klenov, H.,** Abundant amounts of diadenosine 5′,5‴-P^1P^4-tetraphosphate are present and releaseable, but metabolically in active, in human platelets, *Biochem. J.*, 208, 737, 1982.
36. **Dean, B. M. and Perrett, D.,** Studies on adenine and adenosine metabolism by intact human erythrocytes using high performance liquid chromatography, *Biochem. Biophys. Acta*, 437, 1, 1976.
37. **Atkinson, D. E.,** The energy charge of the adenylate pool as a regulatory parameter. Interaction with feedback modifier, *Biochemistry*, 7, 4030, 1968.
38. **Dean, B. M., Perrett, D., and Sensi, M.,** Changes in nucleotide concentrations in the erythrocytes of man, rabbit and rat during short-term storage, *Biochem. Biophys. Res. Commun.*, 80, 147, 1978.
39. **Ericson, A. and De Verdier, C.-H.,** Specimen handling for the assay of adenylates and glycerate 2,3-bisphosphate in erthrocytes, *Scand. J. Clin. Lab. Invest.*, 41, 361, 1981.
40. **Brenton, D. P., Astrin, K. H., Cruikshank, M. K., and Seegmiller, J. E.,** Measurement of free nucleotides in cultured human lymphoid cells using high pressure liquid chromatography, *Biochem. Med.*, 17, 231, 1977.
41. **Yoshika, A., Tanaka, K., Wataya, Y., and Hayatsu, H.,** dATP content in commercial ATP samples, *Chem. Pharm. Bull.*, 30, 2651, 1982.
42. **Tanaka, K., Yoshika, A., Tanaka, S., and Wataya, Y.,** An improved method for the quantitative determination of deoxynucleoside triphosphates in cell extracts, *Anal. Biochem.*, 139, 35, 1984.
43. **Shmukler, H. W.,** Purification of KH_2PO_4 for use as a carrier buffer in ultra-sensitive liquid chromatography, *J. Chromatogr. Sci.*, 8, 581, 1970.
44. **Perrett, D.,** Rapid anion-exchange chromatography of nucleotides in physiological fluids, *Chromatographia*, 16, 211, 1982.
45. **Liebes, L. F., Kuo, S., Krigel, R., Pelle, E., and Silber, R.,** Identification and quantitation of ascorbic acid in extracts of human lymphocytes by HPLC, *Anal. Biochem.*, 118, 53, 1981.
46. **Goday, A., Simmonds, H. A., Webster, D. R., Levinsky, R. J., Watson, A. R., and Hoffbrand, A. V.,** Importance of platelat-free preparations for evaluating lymphocyte nucleotide levels in inherited or acquired immunodeficiency syndromes, *Clin. Sci.*, 65, 635, 1983.
47. **Bartlett, G. R.,** Phosphorus compounds in vertebrate red blood cells, *Am. Zool.*, 20, 103, 1980.
48. **Ritter, E. J. and Bruce, L. M.,** The quantitative determination of deoxyribonucleoside triphosphates using high performance liquid chromatography, *Biochem. Med.*, 21, 16, 1979.
49. **Yeagin, C. D.,** A procedure for the measurement of intracellular deoxyribonucleoside triphosphate pools by thin layer chromatography, *Anal. Biochem.*, 58, 231, 1974.
50. **Pestell, R. Q. W. and Woodland, H. R.,** A procedure for the isolation and determination of deoxyribose derivatives, *Biochem. J.*, 127, 589, 1972.
51. **Garrett, C. and Santi, D. V.,** A rapid and sensitive high pressure liquid chromatography assay for deoxyribonucleoside triphosphates in cell extracts, *Anal. Biochem.*, 99, 268, 1979.
52. **Martinez-Valdez, H., Kothari, R. M., Hershey, H. V., and Taylor, M. W.,** Rapid and reliable method for the analysis of nucleotide pools by reversed phase high performance liquid chromatography, *J. Chromatogr.*, 247, 307, 1982.
53. **Cohen, M. B., Maybaum, J., and Sadee, W.,** Analysis of purine ribonucleotide and deoxyribonucleotide pools in cell extracts by high performance liquid chromatography, *J. Chromatogr.*, 198, 435, 1980.
54. **Maybaum, J., Klein, F. K., and Sadee, W.,** Determination of pyrimidine ribotide and deoxyribotide pools in cultured cells and mouse liver by high performance liquid chromatography, *J. Chromatogr.*, 188, 149, 1980.
55. **Brooker, G.,** Determination of picomole amounts of enzymatically formed adenosine 3′5′-cyclic monophosphate by high pressure anion exchange chromatography, *Anal. Chem.*, 42, 1108, 1970.

56. **Wojcik, W., Olianas, M., Parenti, M., Gentleman, S., and Neff, N. H.,** A simple fluorometric method for cAMP: application to studies of brian adenylate cyclase activity, *J. Cycl. Nucl. Res.*, 7, 27, 1981.
57. **Yamamoto, T., Shimizu, H., Kato, T., and Nagatsu, T.,** Simultaneous determination of guanine nucleotides in rat brain by high-performance liquid chromatography with dual electrochemical detection, *Anal. Biochem.*, 142, 395, 1984.
58. **Plesner, P. and Ottesen, M.,** Determination of diadenosine tetraphosphate in biological materials by high pressure liquid chromatography, *Carlsberg Res. Commun.*, 45, 1, 1980.
59. **Luthe, J. and Ogilivie, A.,** The presence of 5′,5‴-P[1],P[3]-triphosphate (Ap_3A) in human platelets, *Biochem. Biophys. Res. Commun.*, 115, 253, 1983.
60. **Lagosky, P. A. and Chang, F. N.,** The extraction of guanosine 5′-diphosphate, 3′-diphosphate (ppGpp) from *Escherichia coli* using low pH reagents: a reevaluation, *Biochem. Biophys. Res. Commun.*, 84, 1016, 1978.
61. **Kukko, E. I., Kallio, T. M., and Heinonen, J. K.,** A method for concentration of nucleoside triphosphates by co-precipitation with calcium fluoride, *Anal. Biochem.*, 133, 58, 1983.
62. **Goldstein, G.,** Ligand-exchange chromatography of nucleotides, nucleosides and nucleic bases, *Anal. Biochem.*, 20, 477, 1967.
63. **Moran, R. G. and Werkheiser, W. C.,** A boronate column for the separation of ribo- and deoxyribonucleosides, *Anal. Biochem.*, 88, 668, 1978.
64. **Lothrop, C. D. and Uziel, M.,** Rapid preparation of nucleotides from acid-soluble pools by chromatography on silica, as exemplified with acid extracts of cultured cells, *Clin. Chem.*, 26, 1430, 1980.
65. **Bartlett, G. R.,** Phosphorus compounds in the human erythrocyte, *Biochim. Biophys. Acta*, 156, 221, 1968.
66. **Bartlett, G. R.,** Patterns of phosphorus compounds in the red blood cells of man and animals, *Adv. Exp. Med. Biol.*, 6, 245, 1970.
67. **Bartlett, G. R.,** *Red Cell Metabolism; a Review Highlighting Changes During Storage,* Greenwalt, T. J. and Jamieson, G. A., Eds., Grune & Stratton, New York, 1974, 5.
68. **Bartlett, G. R.,** Phosphorus compounds in red cells of the chicken and duck embryo and hatchling, *Comp. Biochem. Physiol.*, 55A, 207, 1976.
69. **Bartlett, G. R.,** Phosphorus compounds in red cells of reptiles, amphibians and fish, *Comp. Biochem. Physiol.*, 55A, 211, 1976.
70. **Bartlett, G. R.,** Phosphorus compounds in reptilian and avian red blood cells: developmental changes, *Comp. Biochem. Physiol.*, 61A, 191, 1978.
71. **Bartlett, G. R.,** Phosphorus assay in column chromatography, *J. Biol. Chem.*, 234, 466, 1959.
72. **Seki, N. and Saito, T.,** Quantitative comparisons of certain nucleotdies in the hepatic tissues of some aquatic animals, *Int. J. Biochem.*, 2, 276, 1971.
73. **Brown, P. R., Agarwal, R. P., Gell, J., and Parks, R. E.,** Nucleotide metabolism in the whole blood of various vertebrate: enzyme levels and the use of high pressure liquid chromatography for the determination of nucleotide patterns, *Comp. Biochem. Physiol.*, 43B, 891, 1972.
74. **Scholar, E. M., Brown, P. R., Parks, R. E., and Calabresi, P.,** Nucleotide profiles of the formed elements of human blood determined by high-pressure liquid chromatography, *Blood,* 41, 927, 1973.
75. **Bode, C., Zelder, C., Rumpelt, H. J., and Wittkamp, U.,** Depletion of liver adenosine phosphates and metabolic effects of intravenous infusion of fructose or sorbitol in man and in the rat. *Eur. J. Clin. Invest.*, 3, 436, 1972.
76. **Reiss, P. D., Zuurendonk, P. F., and Veech, R. L.,** Measurement of tissue purine, pyrimidine and nucleotides by radial compression high performance liquid chromatography, *Anal. Biochem.*, 140, 162, 1984.
77. **Kaminsky, Y. G., Kosenko, E. A., and Kondrashova, M. N.,** Alteration of adenine nucleotides pool in old rat liver and its normalisation with ammonium succinate, *FEBS Letts.*, 159, 259, 1983.
78. **Pogson, C. I., Gurnah, S. V., and Smith, S. A.,** A sensitive and specific assay for GTP and GDP in tissue extracts, *Int. J. Biochem.*, 10, 995, 1979.
79. **Durie, D. J. B., Adam, K., Oswald, I., and Flynn, I. W.,** Sleep: cellular energy charge and protein synthesis capability, *IRCS Med. Sci.*, 6, 351, 1978.
80. **Edwards, R., Young, A., and Wiles, M.,** Needle biopsy of skeletal muscle in the diagnosis of myopathy and the clinical study of muscle function, *N. Engl. J. Med.*, 302, 261, 1980.
81. **Rahim, Z. H. A., Perrett, D., and Griffiths, J. R.,** Skeletal muscle purine nucleotide levels in normal and phosphorylase kinase deficient mice, *FEBS Lett.*, 69, 203, 1976.
82. **Perrett, D. and Rolston, D.,** unpublished results, 1984.
83. **Bartlett, G. R.,** Phosphate compounds in rat erythrocytes and recticulocytes, *Biochem. Biophys. Res. Commun.*, 70, 1055, 1976.
84. **Peters, G. J., De Abreu, R. A., Oosterhof, A., and Veerkamp, J. H.,** Concentration of nucleotides and deoxynucleotides in peripheral and phytohemaglgutinin-stimulated mammalian lymphocytes: effects of adenosine and deoxyadenosine, *Biochim. Biophys. Acta,* 759, 7, 1983.

85. **De Abreu, R. A., Van Baal, J. M., Bakkeren, J. A. J. M., De Bruyn, C. H. M. M., and Schretlen, E. D. A. M.,** High performance liquid chromatographic assay for nucleotides in lymphocytes and malignant lymphoblasts, *J. Chromatogr.*, 227, 45, 1982.
86. **Pross, S. H., Klein, T. W., and Fishel, C. W.,** Ribonucleoside and nucleotide components of normal human blood lymphocytes and platelets, *Proc. Soc. Exp. Biol. Med.*, 154, 508, 1977.
87. **D'Souza, L. and Glueck, H. I.,** Measurement of nucleotide pools in platelets using high pressure liquid chromatography, *Thromb. Haemostas. (Stuttgart)*, 38, 990, 1977.
88. **Burnette, B., McFarland, C. R., and Batra, P.,** Rapid, isocratic separation of purine nucleotides using strong, anion exchange high performance liquid chromatography, *J. Chromatogr.*, 277, 137, 1983.
89. **Taylor, M. W., Kothari, R. M., Holland, G. D., Martinez-Valdez, H., and Zeige, G.,** A comparison of purine and pyrimidine pools in Bloom's syndrome and normal cells, *Cancer Biochem. Biophys.*, 7, 19, 1983.
90. **Nissinen, E.,** Analysis of purine and pyrimidine bases, ribonucleosides and ribonucleotides by high pressure liquid chromatography, *Anal. Biochem.*, 106, 497, 1980.
91. **Shapiro, I. M., Golub, E. E., May, M., and Rabinowitz, J. L.,** Studies of nucleotides in growth-plate cartilage: evidence linking changes in cellular metabolism with cartilage calcification, *Biosci. Rep.*, 3, 345, 1983.
92. **Leese, H. J., Bigger, J. D., Mroz, E. A., and Lechene, C.,** Nucleotides in a single mammalian ovum or preimplantation embryo, *Anal. Biochem.*, 140, 443, 1984.
93. **Heldt, H. W., Portis, A. R., McC.Lilley, R., Mosbach, A., and Chon, C. J.,** Assay of nucleotides and other phosphate-containing compounds in isolated chloroplasts by ion exchange chromatography, *Anal. Biochem.*, 101, 278, 1980.
94. **Standard, S. A., Perrett, D., and Bray, C. M.,** Nucleotide levels and loss of vigour and viability in germinating wheat embryos, *J. Exp. Bot.*, 34, 1047, 1983.
95. **Perrett, D.,** unpublished observations.

PURIFICATION OF PLANT NUCLEOTIDE EXTRACTS PRIOR TO HPLC ANALYSIS

Eric G. Brown

In using HPLC for nucleotide analysis, evaluation of the need for preliminary purification of the tissue extract has to begin with consideration of (1) the nature of the extractant used and (2) the type of tissue to be examined. For the extraction of free nucleotides from the tissue of plants, especially higher plants, most methods employ cold acid extractants which also serve as protein precipitants and enzyme suppressants. The procedure usually involves mechanical disruption of the biological sample in the presence of the extractant and removal of cell debris by centrifuging or filtering. The most commonly employed extractants are dilute aqueous solutions of perchloric acid (PCA) or trichloracetic acid (TCA). In some of the earlier work with plant tissues, ethanol was employed, but, as pointed out by Bieleski,[1] even boiling alcohols do not eliminate the nonspecific phosphatase activity of plant extracts. This activity not only results in loss, through hydrolysis, of endogenous nucleotides, but also exhibits a phosphotransferase function causing the appearance of spurious chromatographic peaks and other additions to the analytical data. Furthermore, significant phosphatase activity can still be observed at temperatures as low as $-28°C$, thus affording the possibility of serious port-mortem changes in the nucleotide composition of tissues or extracts stored in a freezer for any length of time.[1] The selection of the best available extractant is thus an essential prerequisite to ensuring that the extract is truly representative of the in vivo tissue nucleotide pool. Although often used as an extractant in nucleotide studies by HPLC, TCA is not generally to be recommended, since it can give rise to chromatographic artefacts. For example, it has recently been reported[2] that extraction with TCA particularly affects quantitative determination of GDP, CDP, and UDP. Lush et al.[3] described the use of TCA (10% w/v) to extract the nucleotides of mammalian muscle, but advised the addition of methanol (20% v/v) to eliminate the nucleotide interconversion that otherwise occurs when TCA is used alone.

In connection with the special problems that arise in extracting nucleotides from higher plant tissue, particularly in minimizing changes brought about in extracts by the residual, but highly active, phosphatase activity mentioned above, Bieleski[1] has recommended a formic acid-based extractant consisting of an ice-cold monophasic mixture of MeOH-$CHCl_3$-7*M* formic acid (12:5:3 by volume).

The first step in the purification of extracts for nucleotide analysis by HPLC is removal of the extractant, leaving the nucleotides in an aqueous solution of appropriate pH and of low ionic strength. Cold PCA extracts are thus often neutralized with KOH and the precipitated $KClO_4$ removed by centrifuging. This technique suffers from the disadvantage that, although $KClO_4$ has a low solubility in cold water, it is not completely insoluble. Subsequent concentration of the neutralized, centrifuged extract often results in a further crop of crystals. A second problem in neutralization with KOH is that there can be some small loss of nucleotides by occlusion in the precipitate. Both problems are, however, obviated by Khym's procedure[4] which removes PCA by neutralization with a solution (2 volumes) of 0.05 *M* tri-*n*-octylamine in Freon® (1,1,2-trichloro-1,2,2-trifluoromethane). The nucleotides and PCA are, in effect, partitioned between the aqueous and organic phases, and PCA is removed in the organic layer. Working with the protozoan *Tetrahymena pyriformis,* Reinhart and Koroly[5] reported that the small losses of nucleotides observed following neutralization with either KOH or tri-*n*-octylamine could be overcome by adding dicarbonic acid diethyl ester to the acid extract before neutralizing. According to Chen et al.,[6] the most critical condition for the successful use of the tri-*n*-octylamine procedure is the length of time that the tri-*n*-

octylamine/Freon® solution is stored before use. Within 6 days of storage, quantitative results are obtained, but after 24 days there are extensive differences in the results. ATP is especially badly affected, *circa* 60% being dephosphorylated.[6] Where TCA has been used as extractant, most investigators have subsequently removed it by shaking with diethyl ether. It is also possible to use the amine/Freon® technique in this connection. With the procedure of Bieleski[1] described above, the MeOH-$CHCl_3$-formic acid extractant is best removed by adding water to bring the composition of the extract to 12:11:10 (by volume). At this point, the extract separates into two phases. The bulk of the chlorophylls and terpenoids are removed in the organic phase, whereas the nucleotides remain in the aqueous phase.

In extracts of higher plants there is the additional problem of the ubiquitous occurrence in higher plants of pigments, quinones, phenolics, and tannins. These compounds are present to a varying extent in all plant extracts prepared by standard procedures. The practical effects of this on nucleotide analysis by HPLC include substantial and irregular background readings and considerable shortening of column life. Anion-exchange packings are particularly badly affected by near-irreversible binding which progressively diminishes column efficiency. Whereas the major adenine nucleotides can be discerned in HPLC analyses of untreated plant extracts, most other nucleotide components cannot. Quantification is impossible under these conditions. Various attempts have been made to overcome the problems; none is, however, entirely satisfactory at this time. In pre-HPLC days, one of the most widely used methods for the preliminary purification of plant extracts before nucleotide analysis by column chromatography was charcoal adsorption. This procedure was successfully used in a number of laboratories (see, e.g., Brown),[7] but had the disadvantage both of being slow and of resulting in recovery losses (10 to 20%) of nucleotides. Although substantial, these losses were reproducible[7] for individual nucleotides, and it was therefore possible to apply correction factors to quantitative analytical data.[7] More recent techniques include treating the extract with insoluble poly-*N*-vinylpyrrolidone (PVP) which removes the bulk of the phenolics.[8] The selective properties of PVP in column chromatography of nucleotides has been discussed by Lerner et al.[9] Optimal conditions for the binding of plant phenolics to this polymer were investigated by Anderson and Sowers.[10] Use of PVP chromatography for the desalting of nucleotide extracts has also been described.[11] A procedure for the purification of plant extracts before HPLC, and which makes use of charcoal adsorption, ion exchange, and PVP, has been used by Nieman et al.[12] Krauss and Reinbothe[13] used ligand-exchange chromatography on Chelex® columns to purify nucleotide extracts from *Euglena gracilis* before HPLC.

A different approach to the problem of purifying plant extracts has been adopted by Redgewell,[14] who fractionated extracts by ion-exchange chromatography on SP- and QAE-Sephadex columns. The rationale was to separate first into classes of compounds (sugars, amino acids, organic acids, and phosphate esters) and then take each group separately for further analysis. Several advantages were reported[14] for this method over the more conventional use of ion-exchange resins and celluloses.

In the author's laboratory, a critical experimental survey has recently been made of the various methods available for preliminary purification of plant nucleotide extracts before HPLC. Reassessment of the charcoal adsorption procedure for possible use in HPLC studies confirmed that reproducible nucleotide recoveries of 80 to 90% could be obtained using charcoal in a batch procedure, i.e., by stirring it with the extract for 20 min at 0 to 4°C and then centrifuging, washing, and eluting. However, allowing for the time taken to purify the charcoal before use,[7] the process was considerably slower than the PVP batch procedure described below. Other techniques surveyed included use of a column of either Sephadex LH-20 (adsorption and partition packing) or of QAE-Sephadex (formate). The former, in conjunction with the methanol-chloroform-formic acid extraction technique, effectively removed the bulk of the phenolics and pigments, while the QAE-Sephadex column did not.

Batch or column treatment of the plant extracts with cellulose powders such as Cell Debris Remover® (Whatman) also proved effective in removing phenolics and pigments, but the recovery of nucleotides was poor (40 to 50%). It was concluded from the studies that the only available methods worthy of further consideration for preliminary purification were PVP treatment and LH-20 column chromatography. Direct comparison of the two was made using a methanol-chloroform-formic acid extract of tomato fruits. This tissue was chosen, as it is known to contain a relatively high concentration of phenolic compounds. The results clearly indicated that the PVP-treated sample was the cleaner of the two and that recoveries were both quantitative and reproducible. Attempts to improve the procedure still further by varying the ratio of PVP to tissue weight led to lower recovery of the nucleotide content. From the findings it was concluded that 500 mg/10 g tissue (fresh weight) is optimal.

As a result of the foregoing critical evaluation of available procedures for the purification of plant nucleotide extracts prior to HPLC, an improved procedure has been developed[17] and is being successfully used. It utilizes the extraction procedure of Redgewell[14] in which tissue samples (5 to 6 g) are homogenized with 30 mℓ of a monophasic mixture of methanol-chloroform-98% (v/v) formic acid-water (12:5:1:2 by volume). After centrifuging, the pellet is reextracted with 30 mℓ of the same mixture. The clarified supernatants are pooled and the mixture made biphasic by adding water (21 mℓ) and chloroform (16 mℓ). The aqueous layer is collected by centrifuging. Finally, as recommended by Redgewell,[14] the pellet is extracted with methanol-formic acid (20% v/v with respect to methanol and 2% v/v with respect to formic acid). After centrifuging to clarify, the supernatant is added to the original pooled nucleotide extract. Following evaporation *in vacuo* the extract is redissolved in water (6 mℓ) at pH 3. Aliquots of this are taken for batch treatment with Polyclar® AT (insoluble PVP) at the rate of 50 mg/g tissue originally extracted. It was found to be essential to purify the PVP before use by the procedure of Anderson and Sowers.[10] The batch treatment of the plant extract with PVP consists of stirring for 20 min over ice and then centrifuging. Aliquots (1 mℓ) are then taken and purified further by ligand-exchange chromatography on a column of Chelex® 100 (Cu^{2+}, NH_4^+ form).[15] Nucleotides are eluted using the procedure described by Brown et al.[15] for mammalian extracts. However, with the plant extracts it has been found necessary to collect an additional 30 mℓ of the water effluent. After evaporating the combined nucleotide effluent *in vacuo,* the residue was redissolved in a suitable volume of water for HPLC analysis.

Finally, it should be noted that not all high background or poor resolution is attributable to the purity of the extract. Checks also need to be kept on the purity of the eluting buffers. Where phosphate buffers need improvement, Reiss et al.[16] advocate the use of tandem columns of AG1-X8 and Chelex® 100. They also recommend a charcoal column, in line with the others, to remove traces of styrene-divinyl-benzene copolymers that leach from the ion-exchange columns, especially after prolonged use and regeneration.

REFERENCES

1. **Bieleski, R. L.,** The problem of halting enzyme activity when extracting plant tissues, *Anal. Biochem.*, 9, 431, 1964.
2. **Olempska-Beer, Z. and Bautz Freese, E.,** Optimal extraction conditions for high-performance liquid chromatographic determination of nucleotides in yeast, *Anal. Biochem.*, 140, 236, 1984.
3. **Lush, C., Rahim, Z. H. A., Perrett, D., and Griffiths, J. R.,** A microprocedure for extracting tissue nucleotides for analysis by high-performance liquid chromatography, *Anal. Biochem.*, 93, 227, 1979.
4. **Khym, J. X.,** An analytical system for rapid separation of tissue nucleotides at low pressures on conventional ion-exchangers, *Clin. Chem.*, 21, 1245, 1975.

5. **Reinhart, M. P. and Koroly, M. J.,** Analysis of nucleotides from *Tetrahymena* by high-performance liquid chromatography, *Anal. Biochem.*, 119, 392, 1982.
6. **Chen, S.-C., Brown, P. R., and Rosie, D. M.,** Extraction procedures for use prior to HPLC nucleotide analysis using microparticulate chemically bonded packings, *J. Chromatogr. Sci.*, 15, 218, 1977.
7. **Brown, E. G.,** The acid-soluble nucleotides of mature pea seeds, *Biochem. J.*, 85, 633, 1962.
8. **Loomis, W. D. and Battaile, J.,** Plant phenolic compounds and the isolation of plant enzymes, *Phytochemistry*, 5, 423, 1966.
9. **Lerner, J., Dougherty, T. M., and Shepartz, A. I.,** Selective properties of poly-N-vinyl pyrrolidone in column chromatography of nucleotides, their derivatives and related compounds. A preliminary report, *J. Chromatogr.*, 37, 453, 1968.
10. **Anderson, R. A. and Sowers, J. A.,** Optimum conditions for the binding of plant phenols to insoluble PVP, *Phytochemistry*, 7, 293, 1968.
11. **Dougherty, T. M. and Shepartz, A. I.,** Poly-N-vinyl pyrrolidone chromatography. Effect of pH on elution of bases and a proposed mechanism for the adsorption of purines, *J. Chromatogr.*, 43, 397, 1969.
12. **Nieman, R. H., Pap, D. L., and Clark, R. A.,** Rapid purification of plant nucleotide extracts with XAD-2, PVP and charcoal, *J. Chromatogr.*, 161, 137, 1978.
13. **Krauss, G. and Reinbothe, H.,** Purification and fractionation of free nucleotides from *Euglena gracilis* by a combined procedure of ligand-exchange chromatography and anion-exchange gel chromatography, *Anal. Biochem.*, 78, 1, 1977.
14. **Redgewell, R. J.,** Fractionation of plant extracts using ion-exchange Sephadex, *Anal. Biochem.*, 107, 44, 1980.
15. **Brown, E. G., Newton, R. P., and Shaw, N. M.,** Analysis of the free nucleotide pools of mammalian tissues by high-pressure liquid chromatography, *Anal. Biochem.*, 123, 378, 1982.
16. **Reiss, P. D., Zuurendonk, P. F. and Beech, R. L.,** Measurement of tissue purine, pyrimidine, and other nucleotides by radial compression high-performance liquid chromatography, *Anal. Biochem.*, 140, 162, 1984.
17. **Brown, E. G. and Davies, D.,** unpublished.

OPTIMIZATION AND PREDICTION OF RETENTION OF NUCLEIC ACID COMPONENTS SEPARATED BY HPLC

Peter J. M. Van Haastert and Thomas Braumann

INTRODUCTION

An optimized chromatographic separation should provide reproducible data and sufficient resolution of the solutes to be analyzed within the shortest time possible. Obviously, each separation requires the optimization of separation conditions, such as the chromatographic mode, stationary phase, and composition of the mobile phase.[1] In the past, optimizations have usually been carried out by trial and error. The change of chromatographic conditions paralleled the technological and theoretical evolution of HPLC. In 1972 nucleic acid components were separated, in equal measure, by liquid chromatography (LC), thin-layer chromatography (TLC), paper chromatography (PC), and by electrophoresis (Figure 1). At present, almost all separations are carried out by LC. While in 1972 most separations were done by ion-exchange chromatography (IEC) and about a quarter by gel filtration, contemporary LC employs almost exclusively reversed-phase columns. Presently, about three quarters of the articles on the chromatography of nucleic acid components deal with separations by LC, and about 80% of them use HPLC as the technique. Obviously, HPLC on ion exchangers and reversed-phase columns are among the separation modes most suited for the chromatography of nucleic acid components.

HPLC OF NUCLEIC ACID COMPONENTS

Nucleosides and nucleobases are positively charged at acidic pH, uncharged at neutral pH, and negatively charged at basic pH. Therefore, these compounds have been separated on cation exchangers at low pH, on reversed-phase columns at neutral pH, and on anion exchangers at high pH. Cation-exchange chromatography was used initially by Uziel et al.[2] and by Singhal and Cohn.[3] The rapid separation of these compounds by HPLC on a pellicular cation-exchange column was developed by Horváth and Lipsky in 1969.[4] More recently, major and rare tRNA-bases as well as ribo- and deoxyribonucleosides have been separated by cation-exchange chromatography.[5-7]

Reversed-phase liquid chromatography (RPLC) is presently the most commonly used method for the separation of nucleobases and nucleosides.[8-15] Hartwick et al.[8,9] were the first to use this method for the separation of bases and major and modified nucleosides. Several specific assays for nucleobases and nucleosides employ reversed-phase columns, such as the analysis of ribonucleosides in urine,[10] the quantification of the levels of hypoxanthine, thymine, oxopurinol, and thymidine,[11] adenosine and deoxyadenosine,[12] or only adenosine.[13]

Nucleobases and nucleosides are negatively charged at alkaline pH, which makes possible the use of anion exchange chromatography.[16,17]

Nucleotides are negatively charged; thus, they lend themselves to separations on anion exchangers.[18-27] The development of the analysis of nucleotides follows closely the evolution of HPLC theory and technology.[28] The application of IEC to the analysis of nucleic acid components followed directly from the invention of this technique and its application to organic cations by Tompkins et al.[29] IEC was introduced to the field of the nucleic acid components by the pioneering work of Cohn in the early 1950s.[30-32] Anderson et al. developed an automated effluent device,[33] and Samuelson recognized the partition of organic modifier between mobile phase and ion exchanger.[34] Kirkland[35,36] and Horváth et al.[4,37] investigated

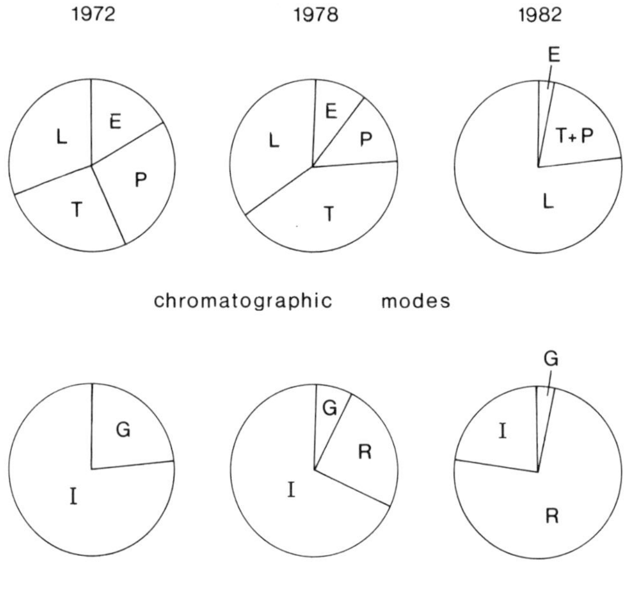

FIGURE 1. The evolution of the chromatography of nucleic acid components between 1972 and 1982. About 150 articles in each year were divided according to the chromatographic mode: L = liquid chromatography, T = thin layer chromatography, P = paper chromatography, and E = electrophoresis. Liquid chromatography was subdivided according to the columns used: I = ion exchanger, G = gel filtration, R = reversed-phase column.

the parameters which affect column efficiency; the development in the late 1960s of pellicular ion-exchange material capable of withstanding high pressures was an important step in the HPLC analysis of nucleic acid components.[37] After the introduction of chemically bonded microparticulate reversed-phase packing materials by Halász and Sebestian,[38] Horváth et al.[39] have used this method for the separation of ribonucleoside monophosphates. Krstulović et al.[40] separated the cyclic nucleotides and Schweinsberg and Loo[41] were able to resolve ATP, ADP, and AMP from other nucleotides, nucleosides, and bases by RPLC.

However, the reversed-phase mode had obvious limitations in the analysis of ionic compounds. Soon after the introduction of ion-pair chromatography, in which a hydrophobic charged molecule is added to the mobile phase and can interact with the hydrophobic stationary phase and with the charged solute, Hoffman and Liao[42] succeeded in separating the mono-, di-, and trinucleotides of cytosine, uracil, guanine, and adenine by RPLC using tetra-*n*-butyl ammonium ions as the ion-pair reagent. Presently, ion-pair RPLC is often used for the separation of nucleotides.[43-48]

The simultaneous analysis of nucleobases, nucleosides, and nucleotides by HPLC has been investigated by many authors.[42,49-60] At present two methods are generally used: (1) anion-exchange chromatography at alkaline pH at which both nucleobases, nucleosides, and nucleotides are negatively charged. The separations require gradient elution with increasing ion concentration and decreasing pH, and are relatively time consuming; (2) RPLC in which the polar nucleotides are retarded by ion pairing with a hydrophobic cation. The separations usually involve gradient elution with increasing content of organic modifier and increasing concentration of polar counterions.

Simultaneous separations of all nucleobases, nucleosides, and nucleotides are rarely carried out; more often, a rapid and complete separation of a particular group of compounds is

Table 1
STEPS IN THE SELECTION OF CHROMATOGRAPHIC CONDITIONS

Description of separation problem
  Number of solutes in sample
  Number of solutes to be identified
  Number of samples to be analyzed
  Physicochemical properties of solutes
Mode of separation
  One sample per run: HPLC, gas chromatography, isotachophoresis
  Multiple samples per run: TLC, electrophoresis, PC
Stationary phase for nucleic acid compounds by HPLC
  Ion-exchanger
  Reversed-phase
  Special columns, e.g., affinity matrix[61]
Elution mode
  Gradient elution, isocratic elution
  Automation
Optimization of mobile phase
  See Table 2

Table 2
STEPS IN THE OPTIMIZATION OF THE MOBILE PHASE COMPOSITION

Selection of fixed and variable parameters
Division of variable parameters in k'-parameters and α-parameters
Vary k'-parameters to get some separation
Vary α-parameters to get optimal resolution
Vary k'-parameters to optimize analysis time

needed.[1,13] In this chapter we will not list all chromatographic conditions available today, but present a general approach to the optimization of the specific chromatographic problems.

OPTIMIZATION OF CHROMATOGRAPHIC SEPARATIONS

The first step in any optimization procedure is the description of the separation problem. This contains a description of (1) the number of solutes which are to be identified, (2) the number of samples to be analyzed, and (3) the physicochemical properties of the solutes such as pK_a values and polarity indices (Table 1). These data are useful for prediction of chromatographic behavior of these compounds. The separation of adenosine and 5'-AMP for the determination of nucleotidase activity can be performed by almost any chromatographic method including TLC. On the other hand, the determination of 2'-AMP, 3'-AMP, and 5'-AMP levels in mixtures containing many other nucleic acid components requires a more sophisticated separation method. Thus, the complexity of the separation problem, the number of samples to be analyzed, and the physicochemical properties lead to the choice of the separation mode and the stationary phase. The variation of the mobile phase components has to result in the optimal separation.

Optimization of Mobile Phase Composition (Table 2)

Mobile phase parameters which can be modified easily are the temperature, pH, ionic strength, nature of the ions, and content of organic modifier. Their effects are different in IEC and RPLC, as will be shown in the next sections. Generally, the mobile phase properties can be divided into two groups: those which affect mainly the retention of all solutes (k'-

Table 3
pK VALUES OF NUCLEIC ACID COMPONENTS

	pK_a	pK_b
Nucleobases		
Adenine	9.8	1, 4.5
Cytosine	12.2	4.45
Guanine	9.6, 12.4	0, 3.2
Hypoxanthine	8.9, 12.1	2.0
Thymine	9.9	—
Uracil	9.5, 13.0	—
Xanthine	7.5, 11.1	0.8
Nucleosides		
Adenosine	12.5	3.5
Cytidine	12.5	4.15
Guanosine	9.2, 12.4	1.6
Inosine	8.8, 12.3	1.2
Thymine	9.8	—
Uridine	8.8	—
Xanthosine	5.7, 13	<2.5

parameters), such as the ionic strength in IEC or organic modifier content in RPLC, and mobile phase components which mainly influence the resolution of solutes (α-parameters), such as pH. A systematic variation of these parameters may help to find quickly the desired separation. One establishes first those parameters which do not affect the separation significantly, such as temperature, ionic strength in RPLC, or organic modifier content in IEC. Then the other parameters are divided into k′- and α-parameters. Subsequently, k′-parameters are varied to obtain retention of the solutes of interest, and the α-parameters are varied to maximize the difference in retention among the solutes. Finally, the k′-parameters are changed to achieve the optimal separation within the shortest time possible. Clearly, it is important to be familiar with the physicochemical properties of the solutes, the chromatographic properties of the stationary phases, and the effects of individual mobile phase components on retardation.

Physicochemical Properties of Nucleic Acid Components

pK Values (Table 3)

Nucleosides and nucleobases are relatively weak bases and weak acids. These compounds are positively charged at pH below their pK_b, neutral between pK_a and pK_b, and negatively charged at pH above pK_a. Therefore, these compounds can be separated on cation and anion exchangers and on reversed-phase columns.

On the other hand, nucleotides are strong acids with pK_a values of approximately 1. The mononucleotides have one negative charge at pH 2.0, the diphosphates have two, and the triphosphates have three negative charges. At pH values above 7.0 the nucleotides gain an additional negative charge at the phosphate, with the exception of the cyclic nucleotides which do not show a secondary dissociation. These compounds can thus be separated by anion-exchange chromatography or by ion-pair RPLC.

Polarity Indices (Table 4)

The polarity of nucleic acid components is important for a prediction of retention in RPLC, but also for IEC. The capacity ratios (k′) of the compounds on a reversed-phase column are a reflection of their polarity indices. A more extensive discussion can be found in the section on RPLC. Adenine is the most hydrophobic nucleobase, while guanine, hypoxanthine, and

Table 4
POLARITY INDICES OF NUCLEIC ACID COMPONENTS

	k′	ln k′
Nucleobases		
Adenine	3.63	1.29
Cytosine	0.45	−0.47
Guanine	1.61	0.48
Hypoxanthine	1.36	0.31
Thymine	1.77	0.57
Uracil	0.75	−0.29
Xanthine	1.26	0.23
Nucleosides and nucleotides		
Adenine	3.63	1.29
Adenosine	4.71	1.55
2′Deoxyadenosine	7.03	1.95
2′-AMP	1.52	0.42
3′-AMP	0.73	−0.31
5′-AMP	0.36	−1.01
3′,5′-cAMP	3.67	1.30
5′-ADP	0.18	−1.70
5′-ATP	0.09	−2.40

Note: The k′ values were calculated from retention data of the compounds on Lichrosorb® RP-18 with 7.5% methanol, 10 m*M* Na H_2PO_4/Na_2HPO_4, pH 7.0 as mobile phase.

xanthine are more polar. Thymine is a relatively hydrophobic pyrimidine; uracil and especially cytosine are very polar (e.g., cytosine is more polar than some adenine nucleotides).

Nucleosides, and especially deoxynucleosides, are less polar than the corresponding nucleobases. On the other hand, the nucleotides are more polar. 3′,5′-cAMP is a remarkably hydrophobic nucleotide, having approximately the same capacity ratio as adenine. The position of the phosphate group in mononucleotides determines the polarity of the compound; the order of increasing polarity is 2′-AMP > 3′-AMP > 5′-AMP. The introduction of additional phosphate groups leads to a further increase of polarity; each additional phosphate group leads to a 50% reduction of k′ values.

ION-EXCHANGE CHROMATOGRAPHY (IEC)

Theory of Retention in IEC[16]

The distribution coefficient (D_N) of an amphoteric solute (NH) between an anion exchanger and an aqueous solution containing a counter ion (A^{n-}) can be described by the equation

$$D_N = \sum_i \Delta K_i + \Delta K_e \tag{1}$$

where ΔK_i represents the contribution of one of the distributions other than anion exchange, such as (1) partition of the molecules between the mobile phase and the matrix due to the hydrophilic backbone and the hydrophobic spacer, and (2) exclusion of the positively charged ions.

The ΔK_e term is related to the actual anion-exchange equilibrium between the solute and the counter ion. For a n-valent counter ion and for a monovalent solute, ΔK_e can be expressed by

$$\Delta K_e = \frac{1}{[A^{n-}]^n} \cdot \frac{K_I}{1 + K_N[H^+]} \tag{2}$$

where $[A^{n-}]$ is the counter ion concentration in the mobile phase, K_I is the ion-exchange equilibrium constant (selectivity coefficient), K_N is the formation constant of the acid NH, and $[H^+]$ is the hydrogen concentration in the mobile phase. Because of this strong dependence of ΔK_e on the pH and the counter ion concentration, the ΔK_e term is relatively easy to predict. The contribution of the individual ΔK_i terms to retention of charged compounds in IEC may be small compared to the ΔK_e term. However, since the source and nature of the ΔK_i terms are often unknown, a straightforward optimization process is usually not possible. On the other hand, the simultaneous presence of distribution processes described by the ΔK_i and ΔK_e terms may provide chromatographic conditions which allow the separation of mixtures which are not possible on using conventional ion-exchange columns.

Effects of Mobile Phase Components on Retention in IEC

Separation of ΔK_i and ΔK_e Terms

The contribution of the ΔK_e term to retention of solutes in ion exchange can be reduced to zero by addition of high counter ion concentrations to the mobile phase. This allows the investigation of the contribution of the ΔK_i term to retention. The effect of the ion concentration on the retention of some cyclic nucleotides on an anion exchanger is shown in Figure 2. At high ion concentrations several compounds show significant retention due to the ΔK_i term. This residual retention is correlated with the hydrophobicities of solutes measured by RPLC. This observation agrees well with the structure of this anion exchanger which contains an aliphatic spacer between the silica gel matrix and the tertiary amino group.

The interaction between a solute and a hydrophobic matrix is removed by the addition of organic modifier to the mobile phase. The retention of hydrophobic solutes at high ion concentrations is reduced by addition of moderate concentrations of acetonitrile (Figure 2B), which indicates that the ΔK_i term results from the reversed-phase properties of the anion exchanger.

At high concentrations of organic modifier and high ion concentrations, the solutes are retarded (Figure 2B). Retention is now inversely proportional to the hydrophobicities of the solutes, as is the case in normal phase chromatography. This confirms the existence of free silanol groups on the stationary phase.

Since reversed-phase interactions are insignificant at moderate concentrations of organic modifier, and normal-phase interactions are not yet effective, the retention of charged solutes is proportional to the net charge at this concentration of organic modifier (Figure 2C). Reversed-phase and normal-phase properties of ion exchangers have been observed by many authors.[16,19,34]

Effect of Mobile Phase Components on ΔK_e Term

In Figure 2 it was shown that solutes are retarded by ion-exchange mechanisms when the mobile phase contains a moderate amount of organic modifier. Therefore, the effects of mobile phase components on ion-exchange interactions have been investigated using a mobile phase containing 50% ethanol (Figure 3). The counter ion concentration and the pH are the two mobile phase characteristics which are generally used to affect the retention of charged solutes in IEC.

As predicted by Equation 2 and illustrated in Figure 3A, ion-exchange interactions are decreased by increasing counter ion concentrations. This figure also demonstrates that the

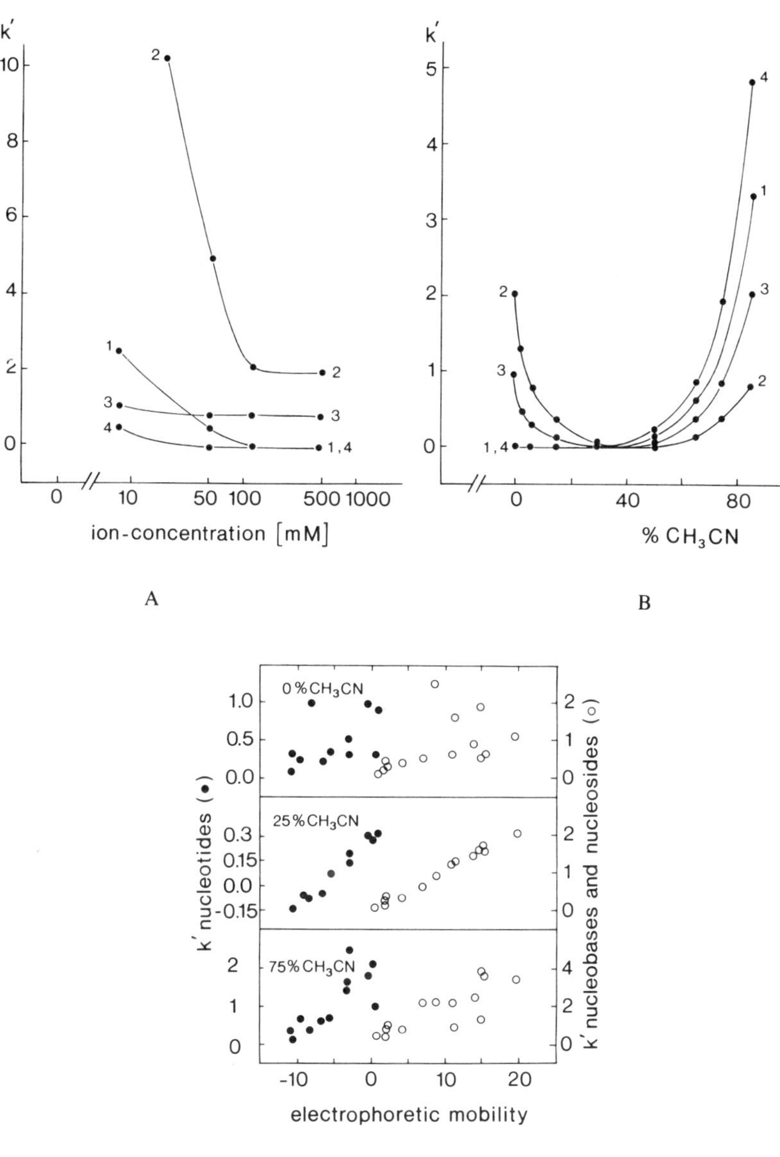

FIGURE 2. Demonstration of reversed-phase and normal-phase characteristics of ion exchangers. (A) Retention of four solutes on the anion exchanger Partisil® SAX at different concentrations of KH_2PO_4/H_3PO_4, pH 3.0. Solutes: 1 = cAMP; 2 = N^6-monobutyryl-cAMP; 3 = 8-benzylamino-cAMP; 4 = 8-amino-cAMP. At high ion concentrations the lipophilic compounds 2 and 3 remain partially retarded by the column; (B) effect of acetonitrile on retention of the same solutes on the anion exchanger at 100 m*M* KH_2PO_4/H_3PO_4, pH 3.0. Notice the inversed order of elution at low and high acetonitrile concentrations; (C) relationship between electrophoretic mobility of 26 nucleobases, nucleosides, and nucleotides and the retention by the cation exchanger Partisil® SCX at 5m*M* triethylammonium formiate, pH 3.0 in (1) the absence of acetonitrile, at (2) 25% acetonitrile, or (3) 75% acetonitrile. (From VanHaastert, P. J. M., *J. Chromatogr.*, 210, 241, 1981. With permission.)

column selectivity is not significantly modified at different counter ion concentrations. Thus, in conventional IEC the ion concentration is a k'-parameter, which affects the retardation of all solutes, rather than the selectivity.

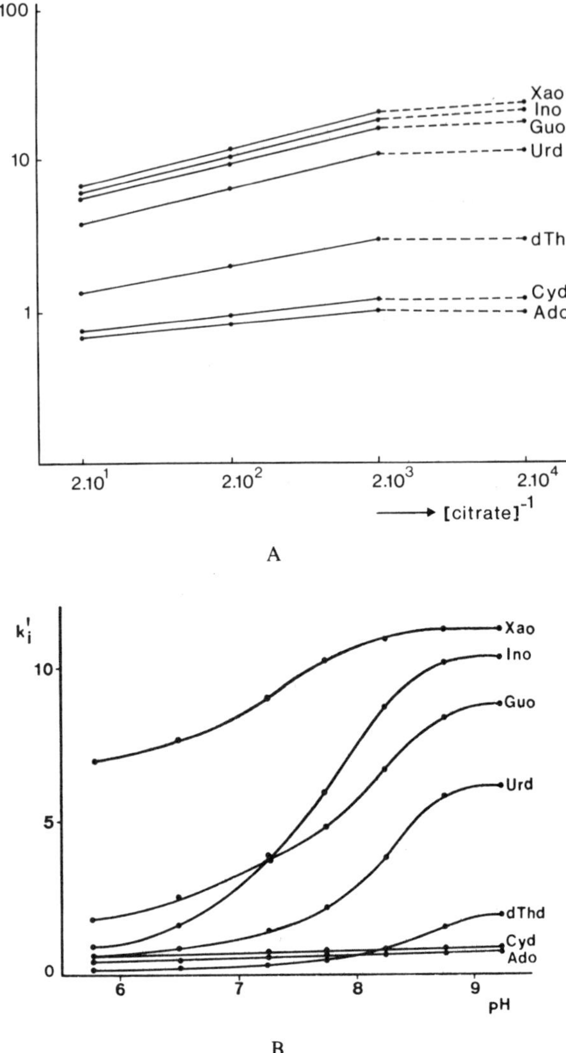

FIGURE 3. The effect of ion concentration (A) and pH (B) on retention of some nucleosides by conventional anion-exchange chromatography. Reversed-phase and normal-phase interactions have largely been suppressed by addition of ethanol. Stationary phase: Aminex® A-28; mobile phase A: 5×10^{-2} to 5×10^{-5} *M* citrate, 5×10^{-3} *M* phosphate buffer, pH 9.0, 50% ethanol; B: 5×10^{-3} *M* citrate, 5×10^{-2} *M* phosphate, 50% ethanol. (From Eksteen, R., Kraak, J. C., and Linssen, P., *J. Chromatogr.*, 148, 413, 1978. With permission.)

The type of the counter ion can influence the ion-exchange equilibrium constant K_l and, therefore, the ΔK_e term. It may also influence the nature of the matrix and, consequently, the ΔK_i term. It has been shown that selectivity in anion-exchange chromatography is not altered significantly by different polar counter ions such as sulfate, chloride, and phosphate, or by different hydrophobic counter ions such as ethanolammonium, ammonium, and tributylammonium in cation-exchange chromatography.[16,63]

Equation 2 predicts the importance of the mobile phase pH in optimizing a separation in IEC. The effect of pH is complex and affects (1) the charge of the solutes according to their pK values, (2) the charge of the counter ions and therefore the effective ion concentration in the mobile phase, and (3) the polarity of the stationary phase. The latter effect can be generally ignored, because the functional groups in cation and anion exchangers are usually strong bases and strong acids, respectively, with pK values outside the pH range of interest (pH 2 to 10). The effect of pH on the charge of the counter ions can either be predicted from the known pK values of the counter ion, or it can be eliminated by selecting a counter ion with a pK value outside the pH range of interest. At the conditions of constant ion concentration in the mobile phase and ion-exchange interaction, the retention behavior of nucleosides and nucleobases at different pH values corresponds closely to predictions made from the pK_b values (Figure 3B).

Effects of Mobile Phase Components on ΔK_i Term

In the previous section it has been shown that the ΔK_i term represents mainly the reversed-phase interactions in the absence of organic modifier, and normal-phase interactions at high concentrations of organic modifier. The concentration of counter ion (above about 10 m*M*) has a negligible effect on reversed-phase and normal-phase interactions (see next section). The pH may have a pronounced effect, sometimes contradictory to its effect on ion-exchange properties. Protonation of a basic group results in increased positive charge and thus in increased polarity. Cation-exchange interactions are increased and reversed-phase interactions are reduced in the absence of organic modifier, while normal-phase interactions are increased at a high content of organic modifier.

The simultaneous existence of ΔK_i and ΔK_e may result in many unexpected properties of ion exchangers. The polarity of the counter ions has a negligible effect on ΔK_e;[63] however, the existence of reversed-phase properties results in attraction of lipophilic counter ions by the stationary phase and thus in reduced ion-exchange interaction, even if the counter ion concentration in the mobile phase remains the same.[63] Obviously, an optimization procedure for a separation problem by IEC should make allowance for the possible simultaneous presence of ion-exchange, reversed-phase, and normal-phase properties.

The Optimization of Separations in IEC

After the selection of an ion exchanger for the solution of a separation problem, the contribution of other than ion-exchange properties can be eliminated by addition of a moderate amount of an organic modifier (e.g., about 20% [v/v] acetonitrile). Based on the knowledge of pK values of the solutes of interest, a pH of the mobile phase is adjusted to give an optimal charge difference among the solutes to be separated. Then the concentration of the counter ions is varied to achieve a rough separation of the solutes. This separation is subsequently optimized by variation of the pH. If the resolution of solutes is not satisfactory, the organic modifier content can be reduced or increased to introduce the reversed-phase and normal-phase interactions, respectively. Finally, the analysis time is optimized by variation of the ion concentration.

For some separation problems it is advantageous to have a mixture of different chromatographic modes. The optimization of separations in IEC will be illustrated with some examples of simple and complex separation problems.

Examples

The first example is a relatively simple separation problem which requires optimization for rapid and routine analyses of several hundreds of samples. For the determination of the substrate specificity of cyclic nucleotide phosphodiesterases, we aimed to measure the hydrolysis of about 25 cyclic AMP derivatives at the nanomolar level. HPLC was chosen as the separation technique because it is sensitive, fast, and reproducible. cAMP and 5′-AMP are easily separated by RPLC,[64] due to the large differences in their polarities (see Table 4). However, cAMP derivatives exhibit large differences in polarity; the most hydrophobic derivative is retarded about 100 times more than the most polar derivative.[65] Thus, each pair of cAMP and 5′-AMP derivative is easily separated by RPLC, but each pair requires a different content of organic modifier. Therefore, the hydrolysis of the cAMP derivatives was measured on the anion exchanger (Partisil® SAX). Reversed-phase interactions were suppressed by addition of an organic modifier to the mobile phase. A neutral pH was selected, and the ion concentration was modified in such a way that cAMP elutes after about 1.5 column volumes (k′ = 0.5). This allows the separation of cAMP from impurities in the enzyme preparation, which elute in the column-void volume. Then, the pH of the mobile phase was modified to obtain the selectivity between cAMP and 5′-AMP of about 2. It is interesting to notice that this α value requires a slightly acidic pH; at alkaline pH, the charge difference between cAMP and 5′-AMP is too large, resulting in excessive selectivity which would make mandatory the use of gradient elution. Finally, the ion concentration was varied to achieve the desired short analysis time. The final mobile phase composition was 50 m*M* KH_2PO_4, 15% propanol-1, 5% methanol, pH 5.0. All cAMP derivatives eluted with k′ values between 0.5 and 0.75, and all 5′-AMP derivatives had k′ values between 1.5 and 2.0. The analysis time was only 4 min.

The second example describes the complex separation of the nucleobases, nucleosides, and mono-, di-, and trinucleotides of adenine, guanine, cytosine, and uracil by the anion-exchanger Micro Pak® AX-10.[19] In this study, Edelson et al. investigated the effect of acetonitrile on the retardation of nucleic acid components. The nucleobases and nucleosides are hardly retarded or even slightly excluded by the anion exchanger at pH 2.85 in the absence of acetonitrile. The nucleotides are retarded to a different extent; e.g., 5′-AMP is retarded two column volumes. The addition of 80% acetonitrile to the mobile phase results in two effects: (1) the resolution of nucleobases and nucleosides is satisfactory (Figure 4A), and (2) under these conditions, nucleotides are not eluted from the column (5′-AMP, e.g., is retarded at least 60 column volumes). Apparently, the column used by Edelson et al. also exhibited normal-phase properties introduced by acetonitrile. These investigators have also shown that the subsequent decrease of acetonitrile concentration resulted in good resolution of the mononucleotides. Finally, they achieved a complete separation of the mono-, di-, and trinucleotides using a linear gradient of increasing ion concentrations without the acetonitrile (Figure 4B). Based on these results, the following elution program with two gradients can be proposed for the separation of all these nucleic acid components in a single run (Figure 4C). The nucleobases and nucleosides are eluted first with a mobile phase consisting of 80% acetonitrile and 20% 0.01 *M* KH_2PO_4, pH 2.85. After 10 min, a linear gradient is started to remove the acetonitrile and to increase the pH to 4.4 in order to get a better resolution of the mononucleotides from the dinucleotides. During this step the mononucleotides start to elute from the column; their elution order and resolution are difficult to predict, but easy to optimize by alteration of the pH of buffer B. After the complete removal of acetonitrile, a new gradient is started from the low to a high ion concentration and the di- and trinucleotides are subsequently eluted.

The last example of the optimization of IEC demonstrates the power as well as the unpredictability of the column if both reversed-phase and ion-exchange interactions are present. Some years ago, we were interested in tissue levels of *S*-adenosyl-methionine (AdeMet). The obvious choice for the analysis of this compound is cation-exchange HPLC, because AdoMet has several positive charges at acidic pH; a chromatogram of a commercial preparation of AdoMet is shown in Figure 5A. At least five peaks were partially resolved; the first three peaks were identified as *S*-adenosylhomocysteine, 5′-deoxy-5′-thiomethyladenosine, and adenine. The identities of peaks 4 and 5, which represented about 50 and 30%, respectively, of the UV absorbing material, were unknown. The suppression of reversed-phase interactions by addition of an organic modifier resulted in the co-elution of both peaks (Figure 5B). The same effect was observed if cation-exchange interactions were suppressed by high ion concentration (Figure 5C). Compounds 4 and 5 had similar retention times when only reversed-phase or only ion-exchange interactions were effective. This behavior sug-

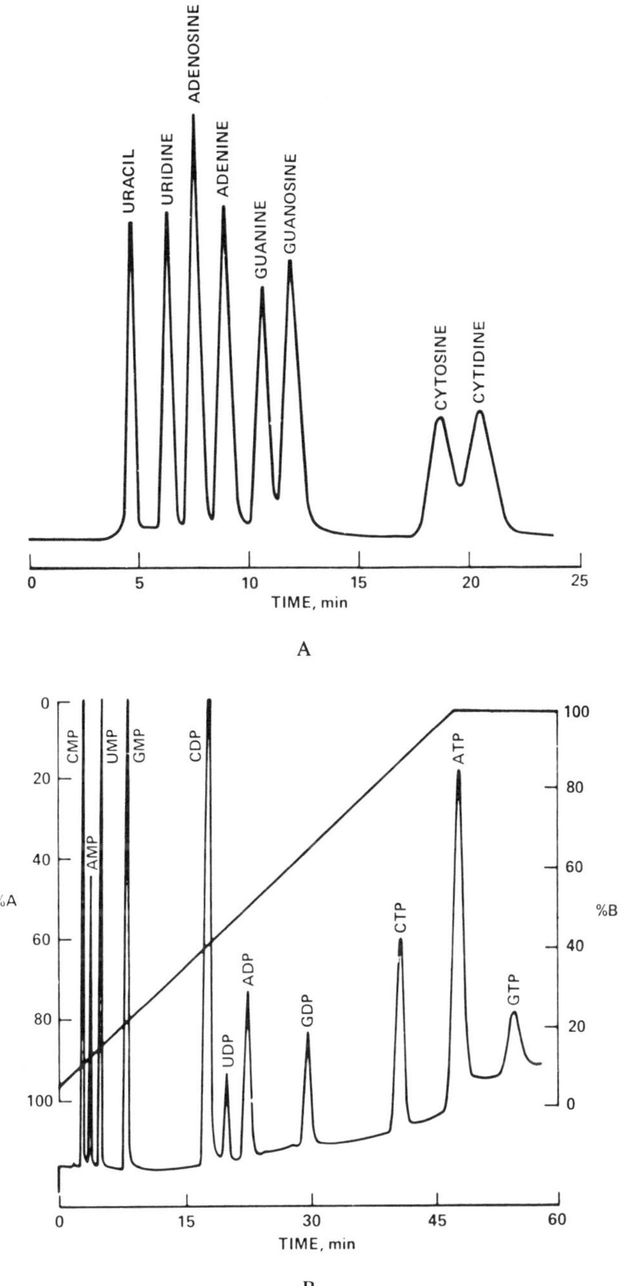

FIGURE 4. Conditions for the simultaneous analysis of nucleobases, nucleosides, and nucleotides of adenine, guanine, uracil, and cytosine by anion-exchange chromatography at acidic pH. (A) Separation of the nucleobases and nucleosides by isocratic elution of the anion-exchanger Micro Pak® AX-10 with 2 m*M* KH_2PO_4, 80% acetonitrile, pH 2.85, at a flow rate of 0.8 mℓ/min; (B) separation of the mono-, di-, and trinucleotides on the same column by elution with a gradient of buffer A (0.01 *M* KH_2PO_4, pH 2.85) and buffer B (0.75 *M* KH_2PO_4, pH 4.4). Flow rate is 2 mℓ/min; (C) proposed elution programs for the simultaneous analysis of nucleobases, nucleosides, and nucleotides by anion-exchange chromatography. Buffer a = 2 m*M* KH_2PO_4, 80% acetonitrile, pH 2.85; buffer b = 0.01 *M* KH_2PO_4, pH 4.4; buffer c = 0.75 *M* KH_2PO_4, pH 4.4. The flow rate is 2 mℓ/min. (From Edelson, E. H., Lawless, J. G., Wehr, C. T., and Abbott, S. R., *J. Chromatogr.*, 174, 409, 1979. With permission.)

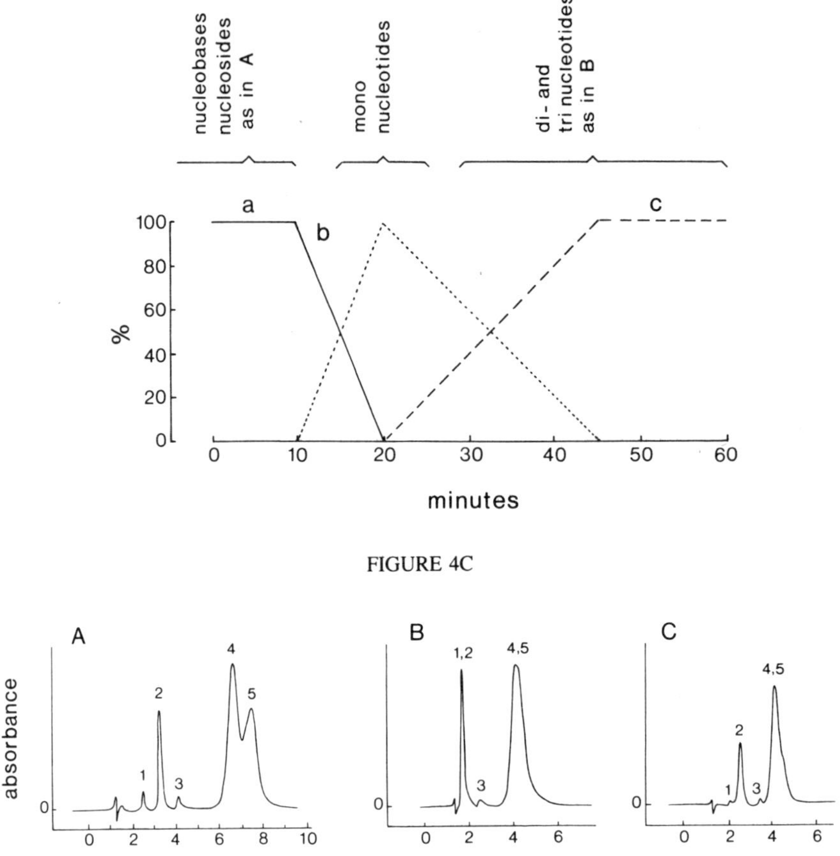

FIGURE 4C

FIGURE 5. Separation of the thio-stereoisomers of *S*-adenosyl-L-methionine and some degradation products by cation-exchange chromatography. Stationary phase; Partisil® SCX; flow rate, 2 mℓ/min. Solutes: 1 = *S*-adenosyl-L-homocysteine; 2 = 5′-deoxy-5′methylthioadenosine; 3 = adenine; 4 = (−) *S*-adenosyl-L-methionine; 5 = (+)*S*-adenosyl-L-methionine. Mobile phase: A = 1 m*M* ammonium acetate, pH 5; B = 1 *M* ammonium acetate, 30% propanol-1, pH 5; C = 1.9 *M* ammonium acetate, pH 6.3.

gested that compounds 4 and 5 were stereoisomers of AdoMet. Subsequently, it was shown that (1) equal amounts of both compounds were present after boiling the sample, (2) isolation of one of the compounds and reinjection gave one peak, (3) the conversion of purified peak 4 to 5 at 100°C had the same half-life as the conversion of peak 5 to 4 (27 min), and (4) both peaks had the same rate of degradation at alkaline pH. Since methionine does not racemize at 100%,[66] the peaks 4 and 5 were the two stereoisomers of AdoMet at the sulfur atom. The natural isomer is (−)-AdoMet, and the tissue extract contained only peak 4.

REVERSED-PHASE LIQUID CHROMATOGRAPHY (RPLC)

Although RPLC was introduced into practice only 10 years ago, chromatography with nonpolar bonded stationary phases and polar (water-containing) mobile phases is now the most important separation technique in LC. The success of RPLC is due to its versatility and ease of operation, long-term stability of RP columns, and the limited number of operational variables for separation optimization. Therefore, it is not surprising that many classical candidates for IEC, including nucleobases, nucleosides, and nucleotides, are now increasingly separated on RP columns.[1,8-10,13,40-48,52,55-57,59,60,64,67-77] However, the incomplete understanding of the retention mechanism of these polar or even charged compounds in RPLC may lead to contradictory results and conclusions. After a short recapitulation of the theory of retention, we attempt to provide rules for prediction of retention of the biologically important nucleic acid components which additionally may help to explain some conflicting results in the literature.

Theory of Retention in RPLC

The retention mechanism of a solute in RPLC, i.e., its distribution between the nonpolar stationary phase and the polar (water-containing) mobile phase can be conveniently described by the solvophobic theory proposed by Horváth and co-workers.[78,79] Accordingly, the driving force for retention is the unfavorable interaction of a solute with the surrounding water molecules of the mobile phase. This leads to a net free energy change upon the association of the solute with the nonpolar ligands of the support. This picture implies that the interaction of the solute with, e.g., the octadecyl-silylated surface of the silica gel, is weak and non-selective.[80] The free energy change is largely determined by the energy which is needed to create a suitably sized cavity within the water molecule network. This hydrophobic (solvophobic) effect[81,82] increases with increasing surface area of the solute, although polar and/or charged substituents may counteract the expulsion from the eluent by introducing Van der Waals and electrostatic interactions between solute and solvent which favor solvation. Conversely, the energy of cavity formation decreases with decreasing surface tension of the eluent. In practice, this is achieved by addition to the mobile phase of an organic modifier with low dielectric constant such as methanol. Although this oversimplified picture of retention in RPLC does not take into consideration some problems and limitations of the solvophobic theory,[83-85] it nevertheless explains the sometimes confusing phenomena observed during separation of nucleobases, nucleosides, and nucleotides.

It has to be mentioned, however, that the stationary phase may exert selective effects upon retention apart from being a "passive" hydrophobic surface. Unreacted silanol groups at the silica gel surface offer additional polar and, above a pH of about 5.0,[86] also anionic sites which contribute to retention of polar or charged solutes. Indeed, it has been shown that the dissociated silanol groups are responsible for retention of protonated amines on RP-18 columns.[87] This problem is not a serious one for the neutral bases and nucleosides, due to the high water content of the mobile phase usually used during separation. The same is true for the negatively charged nucleotides. If cations have to be chromatographed, residual silanol groups can be effectively masked by the addition of a small amount of, e.g., *N,N*-dimethylamino octane to the mobile phase.[88]

Effect of Mobile Phase Components on Retention

Figure 6 shows the structures of adenine, adenosine, and adenosine 3′,5′-monophosphate (cAMP) which may serve as a simple example for the discussion of structure-retention relationships.

The hydrophobic parts of the molecules consist of the base moiety, the core of the ribose and the 3′- and 5′-oxygen of the cyclic phosphate, although the endocyclic N and O atoms possess at least some hydrogen-bonding activities.[89] These structures are responsible for the hydrophobic effect, i.e., the exclusion from an aqueous mobile phase. The amino- and hydroxyl-substituents counteract the hydrophobic effect due to strong hydrogen bonding with water molecules. The phosphate group of cAMP contributes strongly to solution in the eluent, because the charge-bearing group is highly hydrated and additionally repelled from the nonpolar octadecyl-ligands. Surface tension of the eluent, the degree of ionization, and the presence of buffer ions will therefore determine which of the above-mentioned molecular properties are most pronounced in a given reversed-phase system.

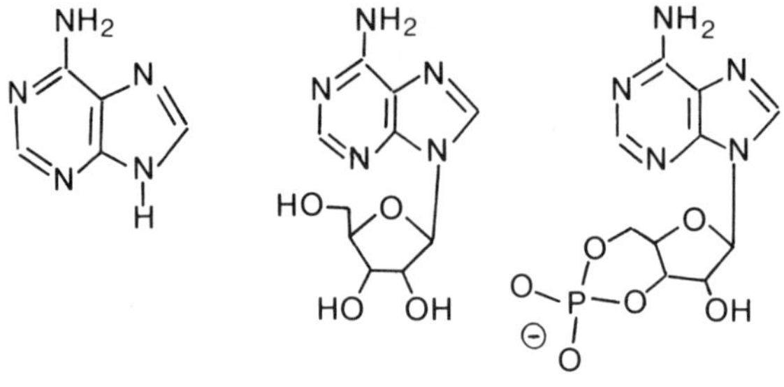

FIGURE 6. Structures of adenine (1), adenosine (2), and cAMP (3) at pH 7.0.

Effect of Organic Modifier on Retention

The hydrophobic effect which determines the extent of retention in RPLC is directly related to the surface tension of the eluent, so that an increase in organic modifier content brings about a decrease in retention. Provided that no secondary equilibria come into play (see below), this relationship can be formulated[90] as:

$$\log k' = S \psi + \log k_w \tag{3}$$

where ψ is the volume fraction of organic solvent in the water-organic solvent mixture, k_w represents the capacity factor of a solute with pure water as mobile phase (usually obtained by extrapolation to the intercept of the ordinate, see Figure 7), and S, the slope of the curve, is related to the solvent strength of the pure organic solvent as well as to the surface area of the solute.[90,91] Figure 7 shows the relationship of Equation 3 for adenosine, adenine, and two different molecular species of cAMP. The species $cAMP^-K^+$ represents a potassium-nucleotide complex which achieves partial electrical neutrality and which is formed at high buffer concentrations in the mobile phase. $cAMP^-$ represents the negatively charged compound and has been obtained using a mobile phase of the same composition, but very low ionic strength. Equation 3 holds only over a limited range of binary composition of the eluent,[92,93] because at low percentages of organic modifier the solvation of the stationary phase changes which is reflected by the curvature of the plots seen in Figure 7. However, a linear relationship between log k′ and ψ can often be used as a good approximation.[91-93] Figure 7 reveals that (1) an effective charge as in $cAMP^-$ reduces retention by a factor of 10, in terms of k_w, (2) neutral compounds or complexes of charged solutes elute in the order of increasing surface area, as long as the water content of the eluent is high and the cavity term of retention is maximal, and (3) an increase in methanol content above 35% yields a reversed sequence of elution; due to the decrease of the surface tension of the eluent to about 50%,[78] the cavity term becomes less important, whereas the polar substituents still contribute to solvation.

Brown and Grushka[69] came to a different conclusion while observing that many nucleosides elute before their corresponding bases at low values of ψ. They took into consideration only the number of polar substituents and explained the superior hydrophobicity of nucleosides by their greater tendency to form dimers through base-stacking interactions which are known to occur in aqueous solution.[94] However, base stacking is strongly concentration dependent and so should be the retention times in a given reversed-phase system. During the experiments performed to obtain Figure 7, we could not observe a concentration dependence of the capacity factor while injecting between 1 pmol and 10 nmol of different nucleic acid components.

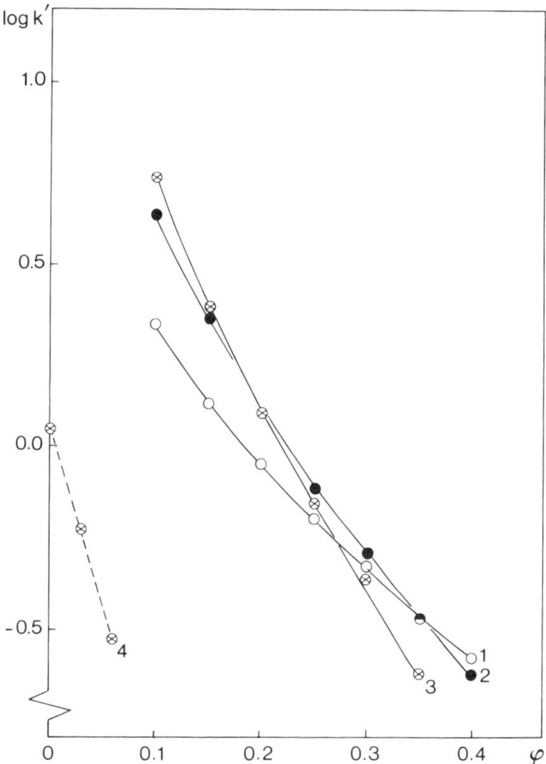

FIGURE 7. Relationship between the logarithm of the capacity factor, k', and the volume fraction of methanol in the mobile phase, ψ. (1) Adenine; (2) adenosine; (3) $cAMP^-K^+$; (4) $cAMP^-$. Aqueous phase was 100 m*M* K^+-phosphate buffer, pH 7.0, except for $cAMP^-$ where 0.5 m*M* K^+-phosphate buffer, pH 7.0 was used.

We have expanded our study to the other major nucleobases, nucleosides, and cyclic nucleotides with respect to the relationship between log k' and ψ (Table 5). Note that the values for the cyclic nucleotides reflect the K^+-complexed species, i.e., they have been obtained using 100 m*M* potassium phosphate buffer, pH 7.0, as part of the aqueous phase. Nucleotides with more than one negative charge are not included, because S and log k_w for these highly polar compounds are difficult to obtain. Table 6 provides two types of information. First, it provides the data necessary for the prediction of retention at any given methanol content of the mobile phase and thus allows the predetermination of resolution for any pair of solutes. Second, log k_w has been proved to be a suitable parameter for modeling quantitatively the hydrophobic nature of many classes of compounds[91,92,95] and can therefore be used in studies of quantitative structure-activity relationships.

Considering the different values of S from Table 5, a general statement has to be modified, i.e., the role of the organic modifier in RPLC is to affect retention of all solutes, but not resolution between a pair of solutes. This statement is only true for compounds of comparable size, shape, and polarity so that, e.g., the group of purine bases shows a similar decrease in retention with an increase in ψ. Solutes of varying size react differently upon changes in ψ, so that the content of organic modifier in the eluent also affects resolution. As an example, adenine, adenosine, and cAMP cannot be separated at ψ = 0.30, but are resolved at ψ = 0.09 or ψ = 0.40; the elution order, however, is reversed (Figure 7).

Table 5
RELATIONSHIP BETWEEN THE VOLUME FRACTION OF METHANOL, ψ, AND LOG k′: LOG k′ = LOG k_w − S ψ

Compound	Log k_w	S	n	r
Adenine	0.61	3.17	6	0.996
Guanine	0.17	3.34	6	0.990
Hypoxanthine	0.15	3.42	5	0.992
Uracil	−0.18	2.85	5	0.991
Cytosine	−0.43	2.56	4	0.983
Adenosine	1.04	4.48	6	0.995
Guanosine	0.63	4.83	6	0.993
Inosine	0.55	4.61	6	0.990
Uridine	0.19	3.93	5	0.991
Cytidine	−0.06	3.53	5	0.986
cAMP	1.26	5.61	6	0.994
cGMP	0.79	5.89	6	0.997
cIMP	0.92	6.01	5	0.982
cUMP	0.53	4.70	5	0.990
cCMP	0.39	4.55	5	0.989

Note: Data were obtained using values of ψ between 0.1 and 0.4, because at methanol concentrations below 0.1 the relationship is nonlinear. The aqueous phase consisted of 100 m*M* K^+ phosphate buffer, pH 6.6. Therefore, the data of the cyclic nucleotides hold for the nucleotide-metal ion complex. n = Number of data points; r = linear correlation coefficient.

Table 6
FUNCTIONAL GROUP VALUES FROM MONOSUBSTITUTED BENZENE DERIVATIVES[a]

Substituent	Methanol-water (τ_M)	Acetonitrile-water (τ_A)	THF-water[b] (τ_T)
CH_3	+0.56	+0.42	+0.41
CH_3CH_2	+1.02	+0.78	+0.48
C_6H_5	+1.73	+1.02	+0.73
OH	−0.82	−0.80	−0.35
CH_2OH	−0.69	−1.06	−0.92
CH_3CH_2OH	−0.36	−0.82	−0.68
Cl	+0.64	+0.47	+0.48
NO_2	−0.13	−0.06	−0.05
CHO	−0.24	−0.44	−0.64

[a] Log k_w for benzene is taken as reference.
[b] THF = Tetrahydrofuran.

Data from Braumann, T., Weber, G., and Grimme, L. H., *J. Chromatogr.*, 261, 329, 1983.

Nature of the Organic Modifier

Among the most commonly employed organic modifiers, methanol and the other lower alcohols exhibit both hydrogen donor and acceptor abilities and will, therefore, be easily incorporated into the network of water molecules. Acetonitrile and tetrahydrofuran, on the other hand, serve only as hydrogen acceptors, thus, changing the structure of the aqueous mobile phase more drastically. Consequently, functional group selectivity is observed in different organic modifiers[91,96] which can be exploited for the separation of structurally related compounds. Table 6 indicates some functional group values, τ_w, which are defined[97] as

$$\tau_w = \log k_{w(j)} - \log k_{w(i)} \tag{4}$$

where k_w represents the (extrapolated) capacity factor in pure aqueous eluent of solutes j and i which differ by a substituent.

Generally, the methanol-water system is the most discriminating eluent with respect to substituent effects upon retention. This is most pronounced for nonpolar groups. Solutes containing aldehyde or short alcohol functions, however, are better separated with, respectively, tetrahydrofuran and acetonitrile. For most practical purposes, methanol is the organic modifier of choice due to its discriminating power, low price, high purity, and relatively low toxicity.

Secondary Equilibria

Since nucleobases, nucleosides, and nucleotides are ionized or at least ionizable compounds, secondary equilibria between the charged solute and an oppositely charged mobile phase additive can greatly affect their retention behavior. Ion-pair formation between solute and pairing ion (hetaeron according to Horváth et al.[98]) results usually in reduction of the electronic charge on the solute molecule with a concomitant increase in retention on nonpolar stationary phase.[98-100] According to the nature of the pairing ion, two different processes have to be taken into consideration.

Metal Binding in the Mobile Phase

Charged nucleic acid components are known to form complexes with many metal ions in aqueous solution.[101] Since metal ions are frequently constituents of the buffered mobile phase used for RPLC of these compounds, complex formation in the eluent may affect retention.

Figure 8 shows a plot of the capacity factor of nucleotides vs. the potassium ion concentration of the phosphate buffer-containing mobile phase. The pH was 7.0 to avoid formation of charges other than the negative charges at the phosphate group. The plot of log k′ vs. log K^+ concentration is of sigmoidal shape for cAMP, indicating no metal-complex formation below 0.1 m*M* K^+ and saturation of metal binding above 50 m*M* K^+. Additionally, the ion pair thus formed achieves partial electrical neutrality, as indicated by a log k′ comparable to that of adenosine. 5′-AMP^{2-} behaves qualitatively in the same manner with respect to metal binding. However, the affinity for K^+ is much smaller, i.e., half-saturation of complex formation occurs at a higher concentration of K^+; the residual polarity of the complex is higher due to the additional charge at the phosphate moiety. ADP^{3-} possesses a high-affinity binding site for K^+,[101] as demonstrated by the increase in log k′ at low K^+ concentration, and may bind a second K^+ at 100 m*M* or above. The uncharged adenosine shows a slight decrease in retention at higher K^+ concentrations, which is related to a change in the properties of the methanol-containing eluent upon addition of buffer components. Included in Figure 8 is a plot of log k′ vs. the Mg^{2+} concentration for $cAMP^-$ (closed circles). Due to the higher affinity of Mg^{2+} to the phosphate, the curve is shifted to lower cation concentrations. These results show that the efficiency of metal binding is dependent on both

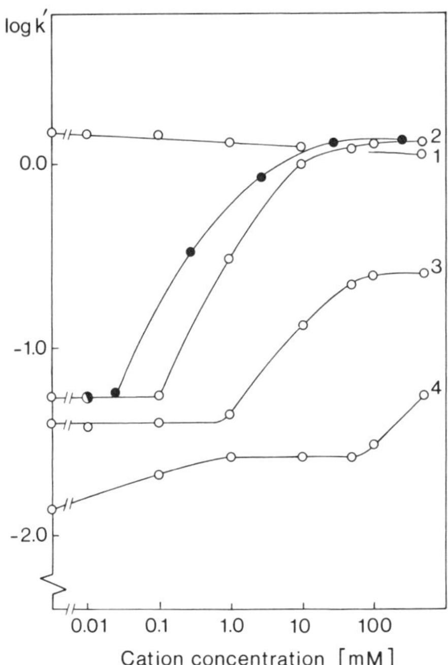

FIGURE 8. Dependence of the logarithm of the capacity factor k', on the cation concentration of the mobile phase. Open circles: K^+; closed circles: Mg^{2+}. (1) Adenosine; (2) cAMP; (3) 3'-AMP; (4) ADP.

the nature of the metal and the structure of the phosphate group.[76] Therefore, the choice of inorganic buffer constituents offers a valuable tool to control selectivity in RPLC. In plain water-methanol eluents, charged compounds interact strongly with the mobile phase and thus possess much shorter retention times than the corresponding bases and nucleosides so that the latter can be conveniently separated. At high ionic strength, the mono- and divalent nucleotides show enhanced retention due to complex formation which can blur the separation from uncharged nucleic acid components. As an example, Krstulović et al.[40] found that adenosine elutes later than cAMP, whereas Anderson and Murphy[52] showed the opposite to be the case. Both groups used RP-18 columns and methanol-phosphate buffer eluents of roughly the same pH. Krstulović et al.,[40] however, chromatographed both compounds at high cation concentration in the eluent, whereas the other group used a mobile phase of low ionic strength. In fact, at a certain ionic strength of the eluent adenosine is not resolved from cAMP (Figure 8).

The di- and triphosphate esters of nucleosides carry at least two negative charges even at high ionic strength; thus, a good resolution of the main nucleotides is difficult to achieve in RPLC with inorganic buffers. These compounds are better separated on ion-exchange columns (see before) or with surface-active pairing ions in the mobile phase (see below).

Surface-Active Pairing Ions

Hydrophobic ions such as tetraalkylammonium compounds or alkylsulphates are increasingly used as the mobile phase additives to control selectivity in ion-pair RPLC.[99,100] Theoretically, it has been argued that these surface-active ions exert their action either via ion pairing in the mobile phase followed by distribution of the ion pair to the nonpolar ligands of the support,[97,99] via a dynamic ion-exchange mechanism,[102] or via a combination of both effects.[100] These hypotheses are no true alternatives, because it has been shown that the

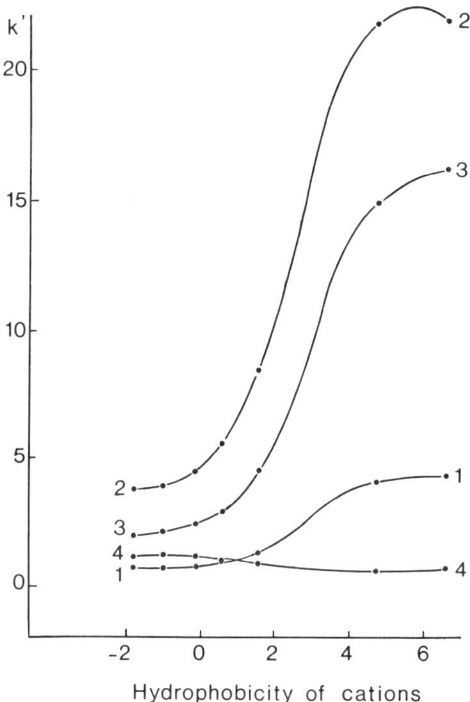

FIGURE 9. Influence of the hydrophobicity of the cations on capacity factors. Cations from left to right: ethanolammonium, ammonium, monoethylammonium, diethylammonium, triethylammonium, tributylammonium, and tetrabutylammonium. For the expression of the hydrophobicity of the cations see Van Haastert.[64] Solutes: 1 = cAMP; 2 = N^6-monobutyryl-cAMP; 3 = 6-chloro-cAMP; 4 = adenosine. (Redrawn from Van Haastert, P. J. M., *J. Chromatogr.*, 210, 229, 1981.)

mechanism of retention is dependent on the actual separation conditions.[100] With relatively small pairing ions, complex formation takes place predominantly in the eluent, whereas large hydrophobic ions are adsorbed onto the stationary phase to form a dynamic ion-exchange system.[102,103] Furthermore, the mechanism of retention may be dependent on the concentration of the pairing reagent.[100] Ion-pair RPLC has been successfully applied to the separation of bases, nucleosides, and nucleotides[22,42,47,48,68] using different chromatographic conditions. In order to obtain improved solute resolution, various chromatographic parameters have to be taken into consideration.

Nature of the pairing ion — Negatively charged nucleotides elute after the corresponding nucleosides[64] if a sufficiently hydrophobic counter ion, e.g., tetrabutylammonium phosphate, is added to the mobile phase. In fact, it has been demonstrated that the solute capacity factor is linearly related to the pairing ion alkyl chain length of large (n = 8 to 14) alkylbenzyl dimethyl ammonium chlorides.[97] Figure 9 indicates a sigmoidal dependence of the capacity factor of the hydrophobicity of smaller alkylammonium compounds which are more often used in ion-pair RPLC of nucleic acid components. A fine adjustment of resolution may be achieved by a mixture of pairing ions of different polarity.[64]

Concentration of pairing ions — A number of studies indicate a complex dependence of the capacity factor upon pairing ion concentration.[97,100,102] At low concentrations (below 0.2 m*M*[97]), depletion in the mobile phase occurs due to adsorption onto the stationary phase. Above 0.5 m*M*, selectivity is, in most cases, independent of the pairing ion concentration.[97]

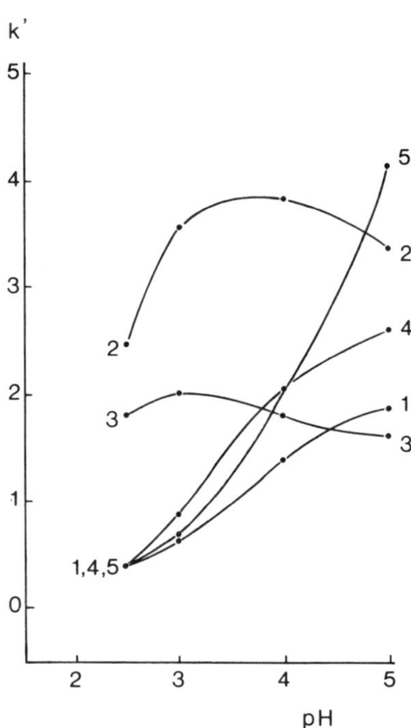

FIGURE 10. Influence of pH on capacity factors. Solutes: 1 = cAMP; 2 = N^6-monobutyryl-cAMP; 3 = 6-chloro-cAMP; 4 = adenosine; 5 = adenine.

However, the hydrophobicity of the pairing ion may influence the concentration dependence of retention.[97,100] In practice, 1 to 10 m*M* of the surface-active ion in the eluent is adequate to obtain good resolution.

Organic modifier — Since in ion-pair HPLC both hydrophobic and electronic forces determine retention, the addition of a low dielectric (e.g., methanol) to the eluent may affect the interaction of charged compounds as well as the hydrophobic effect.[97] Therefore, plots of log k′ vs. the volume fraction of organic modifier are often not linear, as has been described for RPLC with neutral compounds.[91,92,95] From a pragmatic viewpoint, the main effect of the organic modifier is the reduction of surface tension of the eluent with a concomitant decrease in retention of the ion pair; thus, the rules formulated above can be successfully applied also to ion-pair RPLC with hydrophobic pairing ions.

Ionic strength — Buffer ions other than the pairing ion act in a dual way by (1) competing for complex formation with either the solute or the pairing ion and (2) by increasing the surface tension of the mobile phase.[97] These are noticeable, but small, effects on retention and selectivity in ion-pair RPLC,[64,97] so that this parameter is of minor importance for optimizing the mobile phase composition.

pH — The pH of the eluent is the most important mobile phase parameter, because formation or removal of a charge at the solute molecule immediately changes its interaction with pairing ions. Figure 10 shows the influence of the pH of the eluent on capacity factors of different purine derivatives. The adenine-derived compounds show a significant increase in retention between pH 3 and 4, which is directly related to the pK_a of about 3.4 for the N^1 of the base moiety. Substitution of the 6-amino group by a chloro-substituent changes the electron distribution over C^6 and N^1, leading to a nonprotonated compound at relatively low pH;[64] thus, the pH has little effect on retention. Substitution of the N^6-amino group as in N^6-monobutyryl-cAMP decreases the pK_a value of N^1 to approximately 2.5,[64] and shifts the pH-sensitive part of the curve to values below pH 3.0.

Upon addition of pairing ions to the mobile phase, retention of neutral solutes is not affected, whereas solutes with the same charge as the surface-active ion show mostly diminished retention,[64] and solutes with an opposite charge are strongly retarded. Generally, the sequence of elution in ion-pair RPLC is similar to those observed in IEC[68] so that, e.g., AMP elutes before ADP which is less retained than ATP.[42]

Strategies for Optimization of Resolution in RPLC

During the past few years, numerous separation conditions have been described to facilitate the resolution of nucleic acid components in a RPLC system. According to the previous discussion, three different modes of interaction between solute and eluent can be used, i.e., (1) separation on the basis of hydrophobicity alone, (2) separation on the basis of hydrophobicity and charge, and (3) separation on the basis of hydrophobicity and interaction with pairing ions added to the eluent. In this section some examples are given how to exploit these modes of interaction in RPLC.

Separation on the Basis of Hydrophobicity

At pH 6.0, all major bases and nucleosides exist as uncharged solutes (Table 3), so that retention relies exclusively on the hydrophobic effect, i.e., the reduction of the nonpolar surface area exposed to the surrounding water molecules of the eluent. An inspection of the log k_w values of Table 5 reveals that the separation of hypoxanthine from guanine and uridine, and the resolution of guanosine from adenine may present a problem due to their similarities in hydrophobicity in water-rich eluents. This is illustrated in Figure 11 where these compounds are not resolved using 3% methanol in an aqueous eluent.

There are two possibilities to overcome poor resolution as observed in Figure 11: first, by exploitation of differences in values of S, the slope of the plot of log k′ vs. the volume fraction of the organic modifier (Table 5), because molecules with different sizes do not react in the same way upon a given change in mobile phase composition. Figure 12 shows the separation of the same mixture of nucleic acid components using a linear gradient from 100% water to 20% methanol in water within 15 min. As expected, adenine elutes much later than guanosine reflecting the greater sensitivity of nucleosides to an increase in the methanol content compared to the bases. In fact, Hartwick and Brown[8] have demonstrated with a less steep gradient that adenine may elute even after thymidine. Hypoxanthine and guanine, however, are not resolved, because both compounds possess comparable log k_w *and* S values (Table 5). Second, this latter problem can be solved by making use of differences in pK values (Table 1). Introduction of a unit charge reduces the capacity factor of the otherwise unchanged molecule by a factor of 10 (see below).

Provided that all compounds under study are either uncharged or carry the same charge, the following sequence of elution is generally obtained: cytidine < uracil < hypoxanthine, guanine < thymine < adenine. This also holds for the ribosides, deoxyribosides, riboside phosphates, and deoxyriboside phosphates. Experimental details for the separation of more than 80 bases, nucleosides, and nucleotides can be found in the work of Hartwick et al.,[9] including rare bases and cyclic nucleotides. Martinez-Valdez et al.[56] separated the ribo-, deoxyribo-, and cyclic nucleotides from HeLa cells, and De Abreu et al.[57] used a nonlinear gradient to separate bases, nucleosides, and cyclic nucleotides in different biological fluids. Methylation of nucleic acid components increases the hydrophobicity of the solute (Table 6) so that separation on the basis of hydrophobicity is a powerful tool for the analysis of methylated bases and nucleosides.[70,72]

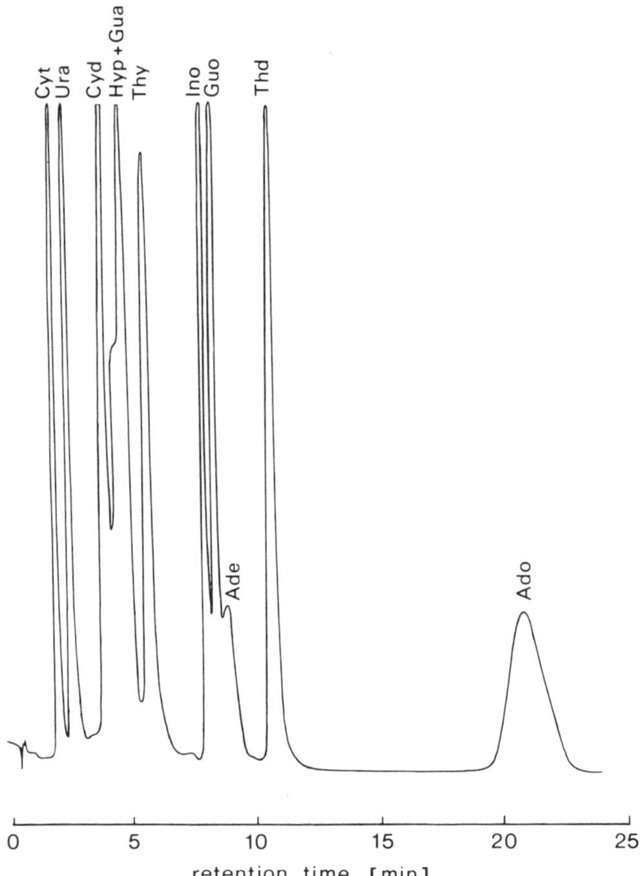

FIGURE 11. Separation of bases and nucleosides on a RP-18 column. Mobile phase was 3% methanol in water (v/v). Solutes: Cyt = cytosine; Ura = uracil; Cyd = cytidine; Hyp = hypoxanthine; Gua = guanine; Thy = thymine; Ino = inosine; Guo = guanosine; Ade = adenine; Thd = thymidine; Ado = adenosine.

Separation on the Basis of Hydrophobicity and Charge

The same solutes used to obtain Figure 11 have been separated in a 10-m*M* phosphate buffer, pH 3.5, containing 3% methanol (Figure 13). Since cytosine, cytidine, guanine, adenine, and adenosine are partly protonated at this pH, elution of these compounds is specifically accelerated and the order of elution of all compounds changes. Note the reversed elution order for cytidine and uracil, the good resolution of guanine from hypoxanthine, and the different position of adenine in the chromatogram. Adenosine, although still being the last-eluted compound, shows a much shorter retention, so that complete resolution of all nucleic acid derivatives is obtained and time of analysis is saved.

In complex samples of biological origin, usually charged nucleotides, as well as bases and nucleosides, occur at the same time. If one is interested in the base and nucleoside profile, the separation from nucleotides is easily achieved using a mobile phase of low ionic strength to avoid complex formation of mono- and divalent nucleotides with buffer cations. Under these conditions, charged compounds elute near the void volume of the column. Figure 14 demonstrates that the addition of 5′-AMP, ADP, ATP, and cAMP to the base/nucleoside sample does not disturb the resolution of uncharged compounds. A similar approach was envisaged by Hartwick and Brown.[8]

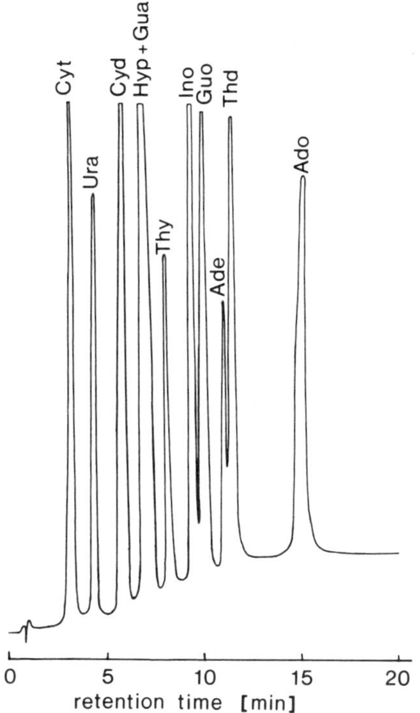

FIGURE 12. Separation of bases and nucleosides on a RP-18 column. Mobile phase was a linear gradient from 100% water to 20% methanol in water (v/v) within 15 min. Abbreviations of solutes are the same as in Figure 11.

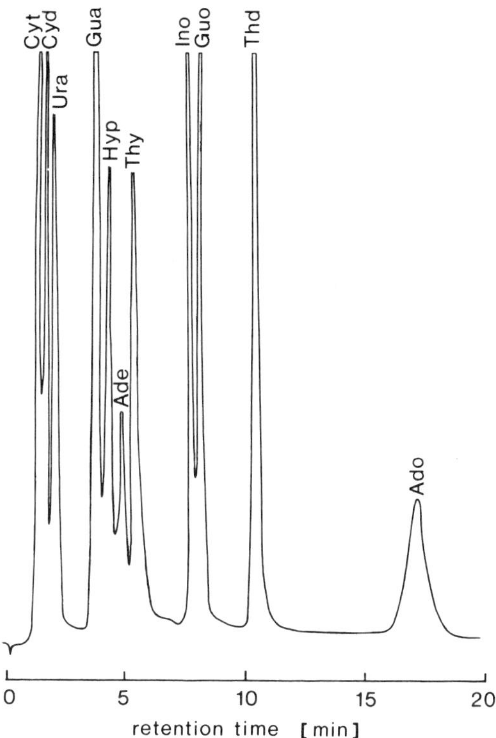

FIGURE 13. Separation of bases and nucleosides on a RP-18 column. Mobile phase was 3% methanol in 10 m*M* K^+ phosphate buffer, pH 3.5 (v/v). Abbreviations of solutes are the same as in Figure 11.

Assenza et al.[15] were able to separate simultaneously cAMP and cGMP, their nucleotides, nucleosides, and bases using an isocratic mobile phase. Although not knowing at that time the influence of cations on complex formation in the mobile phase, the authors optimized the separation conditions to the correct combination of ionic strength and pH of the mobile phase (Figure 15). This is an interesting example due to the combination of ionization and cation binding. The mobile phase consisted of 4 m*M* ammonium phosphate buffer, pH 3.0, so that (1) resolution of hypoxanthine from guanine was managed due to their differences in pK_a values, (2) cyclic nucleotides eluted after their bases, because the negative charge at the phosphate moiety was partially neutralized by ammonium ions, and (3) overlapping of nucleosides and cyclic nucleotides was avoided by using a mobile phase of relatively low ionic strength at which complex formation was not complete and thus the cyclic nucleotides eluted before their corresponding nucleosides (see Figure 8 for the pair adenosine-cAMP). Different cyclic nucleotides have been separated using the same principles in the same report[15] and also in the work of Krstulović et al.[40]

During our studies on uptake and metabolism of cyclic AMP derivatives by neuroblastoma cells, it was of interest to follow the time course of the degradation of cAMP in the serum-containing culture medium. A separation system should be able to separate cAMP from its metabolites 5′-AMP, adenosine, inosine, and hypoxanthine in the shortest time possible. Additionally, cAMP should elute last in order to completely resolve trace amounts of metabolites. RPLC with a mobile phase consisting of 15% methanol in 100 m*M* phosphate buffer, pH 6.6, gave the right order of elution and sufficient selectivity to separate hypoxanthine from inosine (Figure 16). Adenosine, which was not found, would elute just after cAMP under these conditions (see Figure 8).

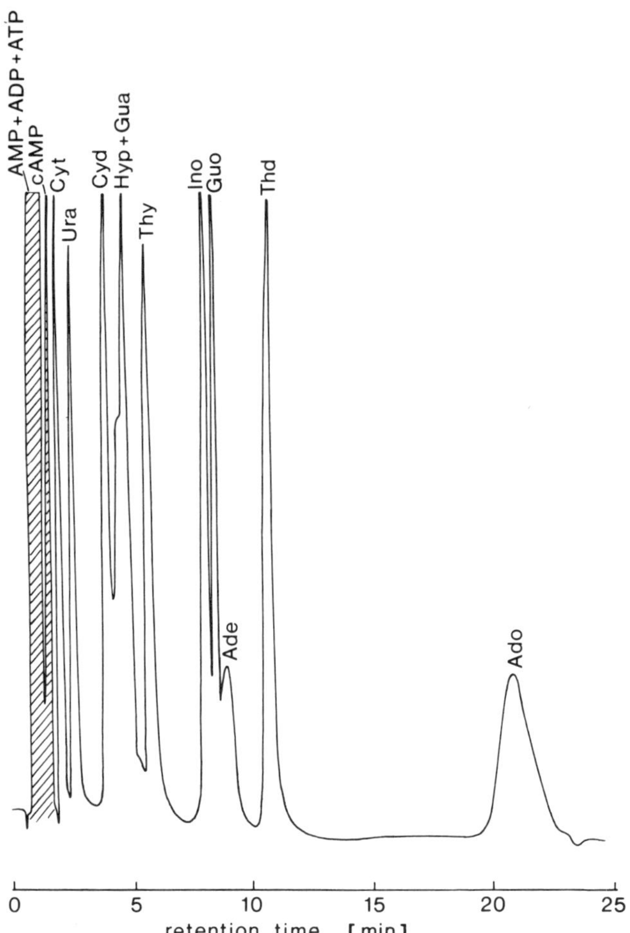

FIGURE 14. Separation of bases, nucleosides, and nucleotides on a RP-18 column. Mobile phase was 3% methanol in water (v/v). Hatched area indicates the charged nucleotides. Abbreviations of solutes are the same as in Figure 11.

Separation on the Basis of Hydrophobicity and Specific Interaction with Pairing Ions

Surface-active ions bind to oppositely charged solutes, either in the mobile phase or at the interface of mobile and stationary phase.[97,102] Thus, retention is dependent both on the net charge of the solute and the hydrophobicity of the ion pair,[97] so that this technique lends itself to full-spectrum recording of nucleic acid components in cell extracts. Within a group of compounds with similar charge, the range of elution follows the rule formulated above for hydrophobic interactions. Examples of separations of mono-, di-, and triphosphate esters of nucleosides are presented in several papers.[22,42,46,47] Differently charged solutes elute in the order of increasing charge so that, for example, the monophosphates are less retained than the diphosphates. The cyclic nucleotides, however, show a deviation from this rule in being more strongly retained than the corresponding noncyclic nucleotides.[22,42]

Using a mobile phase of low methanol concentration containing large tetraalkylammonium cations as the pairing agent, all noncharged bases and nucleosides elute first due to the absence of interaction with the pairing ions, followed by (in this order) the mono-, di-, tri-, and cyclic phosphate esters so that full-spectrum recording is possible. This requires, of course, the use of a mobile phase gradient with increasing concentration of organic modifier, because especially the highly charged nucleotide-tetraalkylammonium complexes

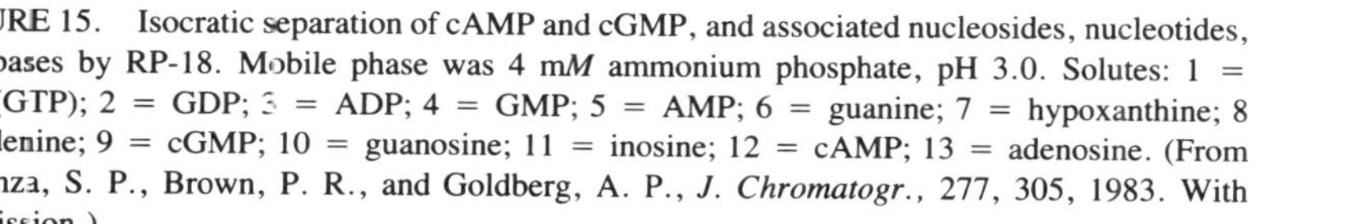

FIGURE 15. Isocratic separation of cAMP and cGMP, and associated nucleosides, nucleotides, and bases by RP-18. Mobile phase was 4 m*M* ammonium phosphate, pH 3.0. Solutes: 1 = ATP(GTP); 2 = GDP; 3 = ADP; 4 = GMP; 5 = AMP; 6 = guanine; 7 = hypoxanthine; 8 = adenine; 9 = cGMP; 10 = guanosine; 11 = inosine; 12 = cAMP; 13 = adenosine. (From Assenza, S. P., Brown, P. R., and Goldberg, A. P., *J. Chromatogr.*, 277, 305, 1983. With permission.)

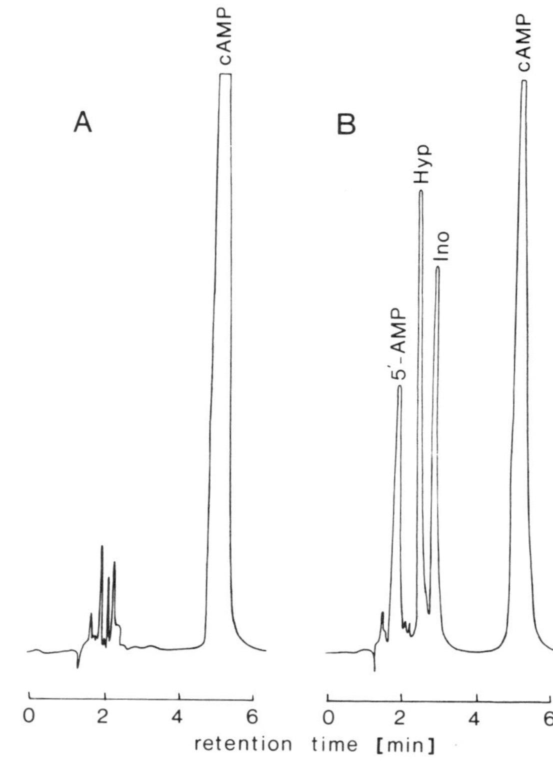

FIGURE 16. Separation by RP-18 of cAMP and its metabolites from the culture medium of neuroblastoma cells containing 10% fetal calf serum. Mobile phase: 15% methanol in 100 m*M* K^+ phosphate buffer, pH 6.6. (A) After addition of cAMP; (B) after 72 hr.

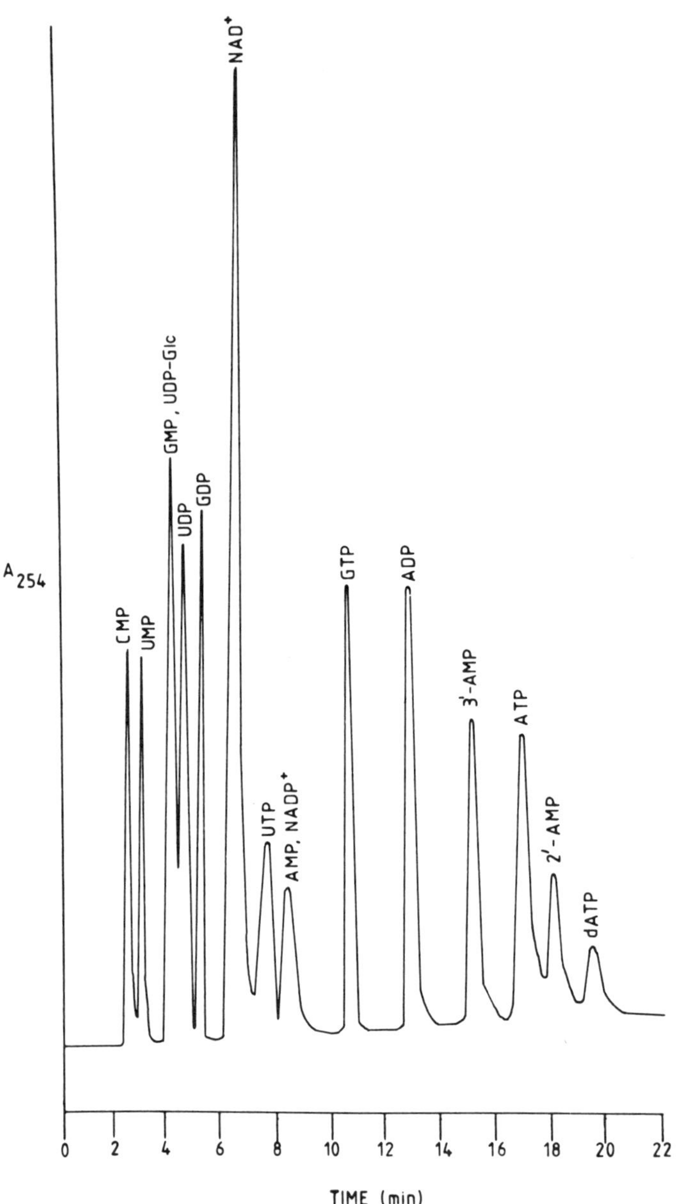

FIGURE 17. Separation of common nucleotides by paired-ion chromatography. The mobile phase was a linear gradient (35 min) of 0 to 30% methanol in 0.1 *M* KH_2PO_4, 5 m*M* tetrabutylammonium phosphate, pH 6.4, at a flow rate of 2 mℓ/min. (From Brown, E. G., Newton, R. P., and Shaw, N. M., *Anal. Biochem.*, 123, 378, 1982. With permission.)

are strongly retained on the column. A good example is presented in Figure 17 where common nucleotides have been separated with tetra-*n*-butylammonium phosphate pairing cations.[22] The same system was successfully applied to the analysis of a liver nucleotide extract. However, complete resolution of nucleotides from bases and nucleosides was not possible (see Figure 4 in Reference 22), because the hydrophobicity of the pairing ion and, therefore, the hydrophobicity of the ion pair were not large enough to achieve sufficiently long retention times for the charged compounds. In this case, tetraalkylammonium compounds with longer alkyl chain length can be employed.[97]

REFERENCES

1. **Zakaria, M. and Brown, P. R.,** High-performance liquid chromatography of nucleotides, nucleosides and bases, *J. Chromatogr.*, 226, 267, 1981.
2. **Uziel, M., Koh, C. K., and Cohn, W. E.,** Rapid ion-exchange chromatographic microanalysis of ultraviolet-absorbing materials and its application to nucleosides, *Anal. Biochem.*, 25, 77, 1968.
3. **Singhal, R. P. and Cohn, W. E.,** Analysis and isolation of nucleosides by ion-exclusion chromatography, *Biochim. Biophys. Acta*, 262, 565, 1972.
4. **Horváth, C. G. and Lipsky, S. R.,** Rapid analysis of ribonucleosides and bases at the picomole level using pellicular cation exchange resin in narrow bore columns, *Anal. Chem.*, 41, 1227, 1969.
5. **Essigman, J. M., Busby, W. F., and Wogan, G. N.,** Quantitative determination of transfer RNA base composition using high-pressure-liquid chromatography, *Anal. Biochem.*, 81, 384, 1977.
6. **Breter, H. J., Seibert, G., and Zahn, R. K.,** Single-step separation of major and rare ribonucleotides and deoxyribonucleotides by high-performance liquid cation-exchange chromatography for the determination of the purity of nucleic acid preparations, *J. Chromatogr.*, 140, 251, 1977.
7. **Herron, D. C. and Shank, R. C.,** Quantitative high-pressure liquid chromatographic analysis of methylated purines in DNA of rats treated with chemical carcinogens, *Anal. Biochem.*, 100, 58, 1979.
8. **Hartwick, R. A. and Brown, P. R.,** Evaluation of microparticle chemically bonded reversed-phase packings in the high-pressure liquid chromatographic analysis of nucleosides and their bases, *J. Chromatogr.*, 126, 679, 1976.
9. **Hartwick, R. A., Assenza, S. P., and Brown, P. R.,** Identification and quantitation of nucleosides, bases and other UV-absorbing compounds in serum, using reversed-phase high-performance liquid chromatography, *J. Chromatogr.*, 186, 647, 1979.
10. **Gehrke, C. W., Kuo, K. C., Davis, G. E., Suits, R. D., Waalkes, T. P., and Borek, E.,** Quantitative high-performance liquid chromatography of nucleosides in biological materials, *J. Chromatogr.*, 150, 455, 1978.
11. **Taylor, G. A., Dady, P. J., and Harrap, K. R.,** Quantitative high-performance liquid chromatography of nucleosides and bases in human plasma, *J. Chromatogr.*, 183, 421, 1980.
12. **Koller, C. A., Stetson, P. L., Nichamin, L. D., and Mitchell, B. S.,** An assay of deoxyadenosine and adenosine in human plasma by HPLC, *Biochem. Med.*, 24, 179, 1980.
13. **Hartwick, R. A. and Brown, P. R.,** Selective analysis for adenosine using reversed-phase high-pressure liquid chromatography, *J. Chromatogr.*, 143, 383, 1977.
14. **Knudson, E. J., Lau, Y. C., Veening, H., and Dayton, D. A.,** Time-concentration studies by high-performance liquid chromatography of metabolites removed during hemodialysis, *Clin. Chem.*, 24, 686, 1978.
15. **Assenza, S. P., Brown, P. R., and Goldberg, A. P.,** Isocratic reversed-phase high-performance liquid chromatographic separation of deoxyribonucleosides and ribonucleosides, *J. Chromatogr.*, 277, 305, 1983.
16. **Eksteen, R., Kraak, J. C., and Linssen, P.,** Conditions for rapid separations of nucleobases and nucleosides by high-pressure anion-exchange chromatography, *J. Chromatogr.*, 148, 413, 1978.
17. **Boháček, J.,** Isocratic separation of DNA bases and deoxyribonucleosides by high-performance liquid chromatography on anex OSTION, *Anal. Biochem.*, 94, 237, 1979.
18. **Brown, P. R.,** The rapid separation of nucleotides in cell extracts by high-pressure liquid chromatography, *J. Chromatogr.*, 52, 257, 1970.
19. **Edelson, E. H., Lawless, J. G., Wehr, C. T., and Abbott, S. R.,** Ion-exchange separation of nucleic acid constituents by high-performance liquid chromatography, *J. Chromatogr.*, 174, 409, 1979.
20. **Ritter, E. J. and Bruce, L. M.,** The quantitative determination of deoxyribonucleoside triphosphates using high performance liquid chromatography, *Biochem. Med.*, 21, 16, 1979.
21. **Cohen, M. B., Maybaum, J., and Sadée, W.,** Analysis of purine ribonucleotides and deoxyribonucleotides in cell extracts by high-performance liquid chromatography, *J. Chromatogr.*, 198, 435, 1980.
22. **Brown, E. G., Newton, R. P., and Shaw, N. M.,** Analysis of the free nucleotide pools of mammalian tissues by high-pressure liquid chromatography, *Anal. Biochem.*, 123, 378, 1982.
23. **Perret, D.,** Rapid anion-exchange chromatography of nucleotides in physiological fluids, *Chromatographia*, 16, 211, 1982.
24. **Burnett, B., McFarland, C. R., and Batra, P.,** Rapid, isocratic separation of purine nucleotides using strong anion-exchange high-performance liquid chromatography, *J. Chromatogr.*, 277, 137, 1983.
25. **Ericson, A., Niklasson, F., and De Verdier, L. H.,** A systematic study of nucleotide analysis of human erythrocytes using an anionic exchanger and high-performance liquid chromatography, *Clin. Chim. Acta*, 127, 47, 1983.
26. **Hsu, D. S. and Chen, S. S.,** Application of anion-exchange column chromatography for determination of alkaline phosphatase activity using 2′-AMP or 3′-AMP as substrate, *J. Chromatogr.*, 276, 169, 1983.
27. **Freese, E., Olempska-Beer, Z., and Eisenberg, M.,** Nucleotide composition of cell extracts analysed by full spectrum recording in high-performance liquid chromatography, *J. Chromatogr.*, 284, 125, 1984.

28. **Cohn, W. E.,** Column chromatography of nucleic acids and their constituents, in *Chromatography,* 2nd ed., Heftman, E., Ed., Reinhold, New York, 1967, chap. 22.
29. **Thompkins, E. R., Khym, J. X., and Cohn, W. E.,** Ion-exchange as a separation mode. I. The separation of fission-produced radioisotopes, including individual rare earths, by complexing elution from Amberlite resin, *J. Am. Chem. Soc.,* 69, 2769, 1947.
30. **Cohn, W. E.,** Separation of mononucleotides by anion-exchange chromatography, *J. Am. Chem. Soc.,* 71, 2275, 1949.
31. **Cohn, W. E.,** The separation of purine and pyrimidine bases and of nucleotides by ion exchange, *Science,* 109, 377, 1949.
32. **Cohn, W. E.,** Heterogeneity in pyrimidine nucleotides, *J. Am. Chem. Soc.,* 72, 2811, 1950.
33. **Anderson, N. G.,** Analytical techniques for cell fractions. II. Spectrophotometric column monitoring system, *Anal. Biochem.,* 4, 269, 1962.
34. **Samuelson, O.,** *Ion Exchange Separations in Analytical Chemistry,* Almqvist and Wiksell, Uppsala, 1963.
35. **Kirkland, J. J.,** A high-performance ultraviolet photometric detector for use with efficient liquid chromatographic columns, *Anal. Chem.,* 40, 391, 1968.
36. **Kirkland, J. J.,** High-speed liquid chromatography with controlled surface porosity supports, *J. Chromatogr. Sci.,* 7, 7, 1969.
37. **Horváth, C. G., Preiss, B. A., and Lipsky, S. R.,** Fast liquid chromatography: an investigation of operating parameters and the separation of nucleotides on pellicular ion exchangers, *Anal. Chem.,* 39, 1422, 1967.
38. **Halász, I. and Sebestian, I.,** Brush type stationary phase with ≡ Si–O–C ≡ bonding in gas chromatography, *Chromatographia,* 7, 371, 1974.
39. **Horváth, C., Melander, W., and Molnár, I.,** Liquid chromatography of ionogenic substances with nonpolar stationary phases, *Anal. Chem.,* 49, 142, 1977.
40. **Krstulović, A. M., Hartwick, R. A., and Brown, P. R.,** Reversed-phase liquid chromatographic separation of 3′,5′-cyclic ribonucleotides, *Clin. Chem.,* 25, 235, 1979.
41. **Schweinsberg, P. D. and Loo, T. L.,** Simultaneous analysis of ATP, ADP, AMP, and other purines in human erythrocytes by high-performance liquid chromatography, *J. Chromatogr.,* 181, 103, 1980.
42. **Hoffman, N. E. and Liao, J. C.,** Reversed phase high performance liquid chromatographic separations of nucleotides in the presence of solvophobic ions, *Anal. Chem.,* 49, 2231, 1977.
43. **Chow, F. K. and Grushka, E.,** High-performance liquid chromatography of nucleotides and nucleosides using outersphere and innersphere metal solute complexes, *J. Chromatogr.,* 185, 361, 1979.
44. **Juengling, E. and Kammermeier, H.,** Rapid assay of adenine nucleotides or creatine compounds in extracts of cardiac tissue by paired-ion reverse-phase high-performance liquid chromatography, *Anal. Biochem.,* 102, 358, 1980.
45. **Knox, J. H. and Jurand, J.,** Zwitterion-pair chromatography of nucleotides and related species, *J. Chromatogr.,* 203, 85, 1981.
46. **Payne, S. M. and Ames, B. N.,** A procedure for rapid extraction and high-pressure liquid chromatographic separation of the nucleotides and other small molecules from bacterial cells, *Anal. Biochem.,* 123, 151, 1982.
47. **Crowther, J. B., Caronia, J. P., and Hartwick, R. A.,** An improved method for the determination of oligonucleotide chain length using phosphodiesterase hydrolysis and ion-pair high-performance liquid chromatography, *Anal. Biochem.,* 124, 65, 1982.
48. **Murray, J. and Thomson, A. B.,** Reverse phase ion pair separation of nucleotides and related products in fish muscle, *J. High Resol. Chromatogr. Chromatogr. Commun.,* 6, 209, 1983.
49. **Anderson, N. B., Green, J. G., Barber, M. L., and Ladd, F. C.,** Analytical techniques for cell fractions. III. Nucleotides and related compounds, *Anal. Biochem.,* 6, 153, 1963.
50. **Goldstein, G.,** Ligand-exchange chromatography for nucleotides, nucleosides and nucleic acid bases, *Anal. Biochem.,* 20, 477, 1967.
51. **Murakami, F., Rokushika, S., and Hatano, H.,** Cation-exchange chromatography of nucleotides, nucleosides and nucleic bases, *J. Chromatogr.,* 53, 584, 1970.
52. **Anderson, F. S. and Murphy, R. C.,** Isocratic separation of some purine nucleotide, nucleoside, and base metabolites from biological extracts by high-performance liquid chromatography, *J. Chromatogr.,* 121, 251, 1976.
53. **Floridi, A., Palmerini, C. A., and Fini, C.,** Simultaneous analysis of bases, nucleosides and nucleoside mono- and polyphosphates by high-performance liquid chromatography, *J. Chromatogr.,* 138, 203, 1977.
54. **Bakay, B., Nisinen, E. A., and Sweetman, L.,** Analysis of radioactive and nonradioactive purine bases, nucleosides, and nucleotides by high-speed chromatography on a single column, *Anal. Biochem.,* 86, 55, 1978.
55. **Wakizaki, A., Kurosaka, K., and Okuhara, E.,** Rapid separation of DNA constituents, bases, nucleosides and nucleotides, under the same chromatographic conditions using high-performance liquid chromatography with a reversed-phase column, *J. Chromatogr.,* 162, 319, 1979.

56. **Martinez-Valdez, H., Kothari, R. M., Hershey, H. V., and Taylor, M. W.,** Rapid and reliable method for the analysis of nucleotide pools by reversed-phase high-performance liquid chromatography, *J. Chromatogr.*, 247, 307, 1982.
57. **De Abreu, R. A., Van Baal, J. M., De Bruyn, C. H. M. M., Bakkeren, J. A. J. M., and Schretlen, E. D. A. M.,** High-performance liquid chromatographic determination of purine and pyrimidine bases, ribonucleosides, deoxyribonucleosides and cyclic ribonucleotides in biological fluids, *J. Chromatogr.*, 229, 67, 1982.
58. **Naikwadi, K. P., Rokushika, S., and Hatano, H.,** Ion chromatography of nucleobases, nucleosides and nucleotides using a dual detection system, *J. Chromatogr.*, 280, 261, 1983.
59. **Zakaria, M., Brown, P. R., and Grushka, E.,** Mechanisms of retention in reversed-phase liquid chromatographic separation of ribonucleotides and ribonucleosides and their bases, *Anal. Chem.*, 55, 457, 1983.
60. **Ramos, D. L. and Schoffstall, A. M.,** Reversed-phase high-performance liquid chromatographic separation of nucleosides and nucleotides, *J. Chromatogr.*, 261, 83, 1983.
61. **Kato, Y., Seita, T., Hashimoto, T., and Shimizu, A.,** Separation of nucleic acid bases by high-performance affinity chromatography, *J. Chromatogr.*, 134, 204, 1977.
62. **Sober, H. A., Ed.,** *Handbook of Biochemistry — Selected Data for Molecular Biology,* 2nd ed., CRC Press, Boca Raton, Fla., 1970, G8.
63. **Van Haastert, P. J. M.,** High-performance liquid chromatography of nucleobases, nucleosides and nucleotides. II. Mobile phase composition for the separation of charged solutes by ion-exchange chromatography, *J. Chromatogr.*, 210, 241, 1981.
64. **Van Haastert, P. J. M.,** High-performance liquid chromatography of nucleobases, nucleosides and nucleotides. I. Mobile phase composition for the separation of charged solutes by reversed-phase chromatography, *J. Chromatogr.*, 210, 229, 1981.
65. **Van Haastert, P. J. M., Dijkgraaf, P. A. M., Konijn, T. M., Garcia Abbad, E., Petridis, G., and Jastorff, B.,** Substrate specificity of cyclic nucleotide phosphodiesterase from beef heart and from *Dictyostelium discoideum, Eur. J. Biochem.*, 131, 659, 1983.
66. **Pike, M. C., Kredich, N. M., and Snyderman, R.,** Requirement of S-adenosyl-L-methionine-mediated methylation for human monocyte chemotaxis, *Proc. Natl. Acad. Sci. U.S.A.*, 75, 3928, 1978.
67. **Montgomery, J. A., Johnston, T. P., Thomas, H. J., Piper, J. R., and Temple, C.,** The use of microparticulate reversed-phase packing in high-pressure liquid chromatography of compounds of biological interest, *Adv. Chromatogr.*, 15, 169, 1977.
68. **Brown, P. R., Hartwick, R. A., and Krstulovic, A. M.,** Current state of the art in the HPLC analysis of free nucleotides, nucleosides, and bases in biological fluids, *Adv. Chromatogr.*, 18, 101, 1980.
69. **Brown, P. R. and Grushka, E.,** Structure-retention relations in the reversed-phase high-performance liquid chromatography of purine and pyrimidine compounds, *Anal. Chem.*, 52, 1210, 1980.
70. **Gehrke, C. W., Kuo, K. C., and Zumwalt, R. W.,** Chromatography of nucleosides, *J. Chromatogr.*, 188, 129, 1980.
71. **Jones, D. P.,** Determination of pyridine dinucleotides in cell extracts by high-performance liquid chromatography, *J. Chromatogr.*, 225, 446, 1981.
72. **Christman, J. K.,** Separation of major and minor deoxyribonucleoside monophosphates by reverse-phase high-performance liquid chromatography: a simple method applicable to quantitation of methylated nucleotides in DNA, *Anal. Biochem.*, 119, 38, 1982.
73. **Assenza, S. P. and Brown, P. R.,** Isocratic reversed-phase high-performance liquid chromatographic separation of deoxyribonucleosides and ribonucleosides, *J. Chromatogr.*, 277, 305, 1983.
74. **Debetto, P. and Bianchi, V.,** Reversed-phase high-performance liquid chromatographic analysis of endogenous purine ribonucleotide pools in BHK monolayer cultures, *J. High Resol. Chromatogr. Chromatogr. Commun.*, 6, 117, 1983.
75. **Preston, M. R.,** Determination of adenine, adenosine and related nucleotides at the low picomole level by reversed-phase high-performance liquid chromatography, *J. Chromatogr.*, 275, 178, 1983.
76. **Braumann, T. and Jastorff, B.,** Physico-chemical characterization of cyclic nucleotides by reversed-phase high-performance liquid chromatography. I. Cation-binding in the mobile phase, *J. Chromatogr.*, 329, 321, 1985.
77. **Braumann, T. and Jastorff, B.,** Physico-chemical characterization of cyclic nucleotides by reversed-phase high-performance liquid chromatography. II. Quantitative determination of the hydrophobicity, *J. Chromatogr.*, 350, 101, 1985.
78. **Horváth, C., Melander, W., and Molnár, I.,** Solvophobic interactions in liquid chromatography with non-polar stationary phases, *J. Chromatogr.*, 125, 129, 1976.
79. **Horváth, C. and Melander, W.,** Liquid chromatography with hydrocarbonaceous bonded phases. Theory and practice of reversed phase chromatography, *J. Chromatogr. Sci.*, 15, 393, 1977.
80. **Karger, B. L., Gant, J. R., Hartkopf, A., and Weiner, P. H.,** Hydrophobic effects in reversed-phase liquid chromatography, *J. Chromatogr.*, 128, 65, 1976.
81. **Sinanoglu, O.,** Solvent effects on molecular associations, in *Molecular Associations in Biology,* Pullman, B., Ed., Academic Press, New York, 1968, 427.

82. **Tanford, C.,** *The Hydrophobic Effect,* 2nd ed., Wiley-Interscience, New York, 1980.
83. **Lochmüller, C. H., Hangac, H. H., and Wilder, D. R.,** The effect of bonded ligand structure on solute retention in reversed-phase high-performance liquid chromatography, *J. Chromatogr. Sci.,* 19, 130, 1981.
84. **Hammers, W. E., Meurs, G. J., and De Ligny, C. L.,** Activity coefficients, interfacial tensions and retention in reversed-phase liquid chromatography on LiChrosorb RP-18 with methanol-water mixtures, *J. Chromatogr.,* 246, 169, 1982.
85. **Martire, D. E. and Boehm, R. E.,** Unified theory of retention and selectivity in liquid chromatography. II. Reversed-phase liquid chromatography with chemically bonded phases, *J. Phys. Chem.,* 87, 1045, 1983.
86. **Weber, S. G. and Tramposch, W. G.,** Cation exchange characteristics of silica-based reversed-phase liquid chromatographic stationary phases, *Anal. Chem.,* 55, 1771, 1983.
87. **Sugden, K., Cox, G. B., and Loscombe, C. R.,** Chromatographic behaviour of basic amino compounds on silica and ODS-silica using aqueous methanol mobile phases, *J. Chromatogr.,* 149, 377, 1978.
88. **Unger, S. H. and Chiang, G. H.,** Octanol-physiological buffer distribution coefficients of lipophilic amines by reversed-phase high-performance liquid chromatography and their correlation with biological activity, *J. Med. Chem.,* 24, 262, 1981.
89. **Jastorff, B.,** The effect of cyclic nucleotide derivatives on cell metabolism, in *Cell Regulation by Intracellular Signals,* Swillens, S. and Dumont, J. E., Eds., Plenum Press, New York, 1982, 195.
90. **Snyder, L. R., Dolan, J. W., and Gant, J. R.,** Gradient elution in high-performance liquid chromatography. I. Theoretical basis for reversed-phase systems, *J. Chromatogr.,* 165, 3, 1979.
91. **Braumann, T., Weber, G., and Grimme, L. H.,** Quantitative structure-activity relationships for herbicides. Reversed-phase liquid chromatographic retention parameter, log k_w, *versus* liquid-liquid partition coefficient as a model of the hydrophobicity of phenylureas, s-triazines and phenoxycarbonic acid derivatives, *J. Chromatogr.,* 261, 329, 1983.
92. **Braumann, T. and Grimme, L. H.,** Determination of hydrophobic parameters for pyridazinone herbicides by liquid-liquid partition and reversed-phase high-performance liquid chromatography, *J. Chromatogr.,* 206, 7, 1981.
93. **Schoenmakers, P. J., Billiet, H. A. H., and De Galan, L.,** Description of solute retention over the full range of mobile phase compositions in reversed-phase liquid chromatography, *J. Chromatogr.,* 282, 107, 1983.
94. **Varughese, K. I., Lu, C. T., and Kartha, G.,** The crystal and molecular structure of cyclic adenosine 3′,5′-monophosphate sodium salt, monoclinic form, *J. Am. Chem. Soc.,* 104, 3398, 1982.
95. **Hafkenscheid, T. L. and Tomlinson, E.,** Observations on capacity factor determination for reversed-phase liquid chromatography with aqueous methanol eluents using the solubility parameter concept model and its derivatives, *J. Chromatogr.,* 264, 47, 1983.
96. **Tanaka, N., Goodell, H., and Karger, B. L.,** The role of organic modifiers on polar group selectivity in reversed-phase liquid chromatography, *J. Chromatogr.,* 158, 233, 1978.
97. **Riley, C. M., Tomlinson, E., and Jefferies, T. M.,** Functional group behaviour in ion-pair reversed-phase high-performance liquid chromatography using surface-active pairing ions, *J. Chromatogr.,* 185, 197, 1979.
98. **Horváth, C., Melander, W., and Nahum, A.,** Measurement of association constants for complexes by reversed-phase high-performance liquid chromatography, *J. Chromatogr.,* 186, 371, 1979.
99. **Horváth, C., Melander, W., Molnár, I., and Molnár, P.,** Enhancement of retention by ion-pair formation in liquid chromatography with non-polar stationary phases, *Anal. Chem.,* 49, 2295, 1977.
100. **Tomlinson, E., Jefferies, T. M., and Riley, C. M.,** Ion-pair high-performance liquid chromatography, *J. Chromatogr.,* 159, 315, 1978.
101. **Tu, A. T. and Heller, M. J.,** Structure and stability of metal-nucleoside phosphate complexes, in *Metal Ions in Biological Systems,* Vol. 1, Sigel, H., Ed., Marcel Dekker, New York, 1974, 1.
102. **Kraak, J. C., Jonker, K. M., and Huber, J. F. K.,** Solvent-generated ion-exchange systems with anionic surfactants for rapid separations of amino acids, *J. Chromatogr.,* 142, 671, 1977.
103. **Bartha, A. and Vigh, G.,** Studies in reversed-phase ion-pair chromatography. I. Adsorption isotherms of tetraalkylammonium ion-pair reagents on LiChrosorb RP-18 in methanol-water elements, *J. Chromatogr.,* 260, 337, 1983.

SEPARATION OF NUCLEOBASES AND NUCLEOSIDES ON SILICA GEL

J. C. Kraak

At present, the use of normal phase adsorption systems for the separation of nucleobases and nucleosides is rather limited. This is not surprising, because reversed-phase chromatography has proven to be eminently suited for the separation of these compounds. However, it has been recently demonstrated that normal-phase liquid chromatography on unmodified silica gel can be a powerful technique for the separation of nucleic acids and analogs.[1-3] The main drawback of normal-phase liquid chromatography is the nonaqueous mobile phase which prohibits the injection of aqueous samples. This is important because most analyses for nucleobases and nucleosides have to be done in aqueous solutions. Therefore, normal-phase adsorption chromatography will certainly not compete with reversed-phase systems, but must be considered as a good alternative when a separation cannot be solved with reversed-phase chromatography. Because of the different interaction mechanisms of solutes with the stationary phase (adsorbent) in normal and reversed-phase systems, different selectivities can be obtained.

MOBILE PHASE SELECTION

The mobile phases commonly used in normal-phase adsorption chromatography can be considered as relatively nonpolar. They cannot be used for polar and ionizable substances like nucleobases and nucleosides because of solubility problems and excessively large retardation. In order to separate nucleobases and nucleosides it has been found that the mobile phase must be rather polar and must also contain a certain amount of pure water or aqueous salt solution.[1-5] The most successful mobile phases for the separation of nucleobases and nucleosides were composed of mixtures of dichloromethane or chloroform-methanol-aqueous solutions of acids and salts.[1-3] The influence of the mobile-phase composition on the retention, solubility, and peak shape was investigated extensively by Ryba et al.[2] and Brugman et al.[3] As can be expected, an increase of the methanol content leads to a decrease of the retention of the solutes. The amount of water which can be dissolved in the mobile phase is limited by the solubility of water in dichloromethane-methanol mixture. Nevertheless, when increasing the water content, the retention decreases up to a certain percentage of water. However, when still more water is added, the retention of the solutes increases markedly again. This effect can be explained by a partial filling of the pores of the silica gel with an aqueous phase which occurs when the mobile phase is almost saturated with water.[6] The presence of the aqueous phase in the pores also allows the solutes to be retained on the basis of a liquid-liquid distribution process.[6,7] When the mobile phase is completely saturated with water, the pores are maximally filled with a stationary liquid and the retention is mainly based on liquid-liquid distribution. Under these conditions, extensive thermostatting of the eluent reservoir and column is necessary.

The presence of an acid and/or salt in the mobile phase seems to have a significant influence on the retention, solubility, and peak shape. This is not surprising, because nucleobases and nucleosides are ionizable substances. Evans et al.[1] and Ryba et al.[2] used ammonium formate buffers of different pH values. The best results were obtained in the pH range of 2.5 to 3. Brugman et al.[3] also tested the influence of pH, but with different types of buffers. They found that the type of buffer has a much larger effect on retention and peak shape than the pH. The addition of acetic and chloroacetic acid buffers gave rise to very assymetric peak shapes, poor solubility of many solutes, and long retention times. These effects were not found with different types of sulfonic acids and with perchloric acid. In particular, a mixture of butanesulfonic acid and butanesulfonate (K^+) was found to be a good choice with respect to peak symmetry, solubility, and degree of retention. On the basis of the experiments with different types of acids, the authors suggest that ion pairs are formed between the added anions in the mobile phase and the more basic nucleobases and nucleosides. This favors the solubility of these compounds in the mobile phase and thus diminishes the interaction with the silica gel (smaller retention times), which improves the peak shape significantly (Figure 1).

STATIONARY-PHASE SELECTION

For the separation of nucleobases and nucleosides, unmodified silica gels with surface areas in the range of 200 to 400 mm^2/g have been tested as stationary phases. The particle size ranged from 5 to 10 μm and spherical as well as irregular particles have been applied.[2,3] Some authors[3] show a slight preference for spherical particles. The number of theoretical plates typically ranged between 6000 to 9000 plates per 25 cm, depending on the type of silica gel. The peak shape and column efficiency are significantly worse on large surface area (400 mm^2/g) silica gels.[2,3] The column efficiency as calculated for the nucleobases and nucleosides is about 20 to 30% lower than that obtained for the same columns with a less polar mobile phase, like dichloromethane or cyclohexane, and neutral test solutes. In all studies the column efficiency of the nucleosides was found to be significantly lower than that obtained with the nucleobases.

Brugman et al.[3] investigated the influence of the type of silica gel (supplier) on the retention behavior of nucleobases and nucleosides. They compared five commercially available silica gels, differing in surface area, particle shape, and diameter. The magnitude of retention on the various types of silica gels seemed to correlate well with the surface area of these packings per unit column volume (m^2/m^3). Only minor selectivity differences were found, which indicates the ease of reproducing the activity of the surface of the silica gel in contrast to the reversed-phase materials. Although the selectivity changes are small, it was found that some pairs of solutes can only be separated on one particular silica gel. For instance, Guo and Cyt can be well separated only on Zorbax-Sil®.

Column-to-column reproducibility has also been investigated by measuring the retention behavior on five different columns filled with silica gel from the same batch. The coefficients of variation for the capacity factors range between 2 and 4%, and between 0.6 and 4.5% for the selectivity factors. The silica gel columns were found to be extremely stable and showed only a few percent change in performance and retention characteristics, after pumping through a few hundred liters of mobile phase. This stability is a definite advantage of unmodified silica gel compared to the chemically modified silica gels.

PRACTICAL UTILITY

Normal-phase systems have been applied to the analysis of nucleobases, nucleosides, and derivatives. Pfadenhauer et al.[4] applied a normal-phase system for the analysis of xanthine and hypoxanthine in plasma, using diethylether-*n*-propanol-aqueous acetic acid mixture as the mobile phase. Elrassi et al.[5] applied a mixture of dichloromethane-isopropanol saturated with water for the separation of some pyrimidines. Under these conditions a stationary liquid phase is generated in the pores of the silica gel so that their phase system can be considered to be a liquid-liquid partition system. Horgan et al.[8] separated adenine, adenosine, and some cytokinins on silica gel using ammoniacal chloroform-methanol-water as mobile phase. No information is given by the authors about the column stability with this basic mobile phase. Evans et al.[1] have described a method for the measurement of pyrimidine bases and nucleosides in body fluids. The silica gel was Lichrosorb® Si 100 and the mobile phase consisted

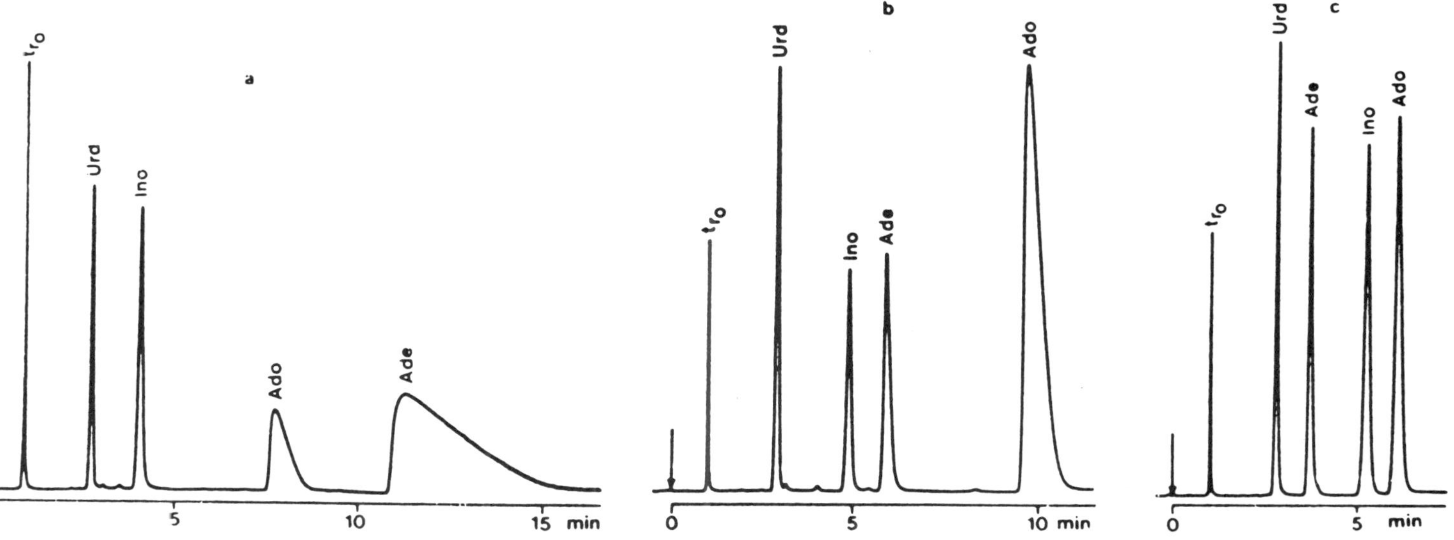

FIGURE 1. Effect of additions of potassium butanesulfonate to the mobile phase on the retention and peak shape of some nucleic acid compounds. Stationary phase: Hypersil®; mobile phase: dichloromethane-methanol-water (835:150:15 v/v) containing 0.05 *M* chloroacetic acid, pH 2.2. (a) No addition; (b) 0.0018 *M* potassium butanesulfonate; (c) 0.01 *M* potassium butane sulfonate. (From Brugman, W. J. Th., Heemstra, S., and Kraak, J. C., *Chromatographia*, 15, 283, 1982. With permission.)

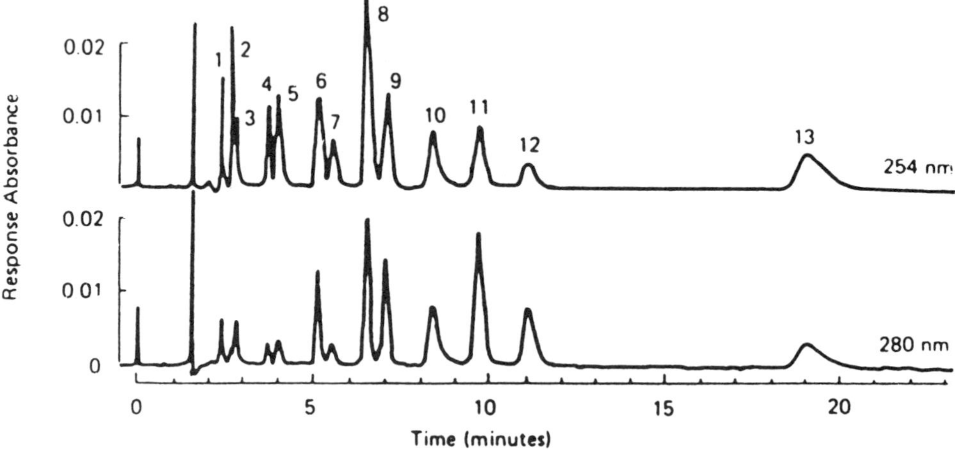

FIGURE 2. Separation of a standard mixture of biologically important pyrimidines. Column; Lichrosorb® Si 100 (250 × 4.6 mm). Mobile phase: dichloromethane-methanol-ammoniumformate buffer (75:22:3). (1) Thymine; (2) uracil; (3) thymidine; (4) 5-hydroxymethyluracil; (5) uridine; (6) 3-methylcytosine; (7) pseudouridine; (8) cytosine; (9) deoxycytidine; (10) 2′-*O*-methylcytidine; (11) cytidine; (12) orotic acid; (13) orotidine. (From Evans, J. E., Tieckelman, H., Naylor, E. W., and Guthrie, R., *J. Chromatogr.*, 163, 29, 1979. With permission.)

of dichloromethane-methanol-aqueous ammonium formate. The isocratic separation of a number of pyrimidines is shown in Figure 2. The authors applied this method to the analysis of uracil and pseudourine in a urine extract. The minimal detectable amount for the investigated pyrimidines ranged between 0.2 and 2 ng, which indicates the applicability of normal-phase systems in trace analysis. Ryba et al.[2] investigated the retention behavior of more than 50 pyrimidine and purine derivatives on the same phase system used by Evans et al. They obtained a complete separation of nucleobases and nucleosides under isocratic conditions in 25 min, as shown in Figure 3. The authors also investigated the sample loadability of the chromatographic system and found it to be rather small (5 to 10 μg). This is related to the solubility of the solutes in the mobile phase.

Brugman et al.[3] also used for the separation of nucleobases and nucleosides a dichloromethane-methanol-water mixture, but containing butanesulfonic acid and potassiumbutanesulfonate. An almost complete separation of these solutes was obtained in 10 min, as can be seen in Figure 4. The potential of this phase system for the analysis of DNA hydrolysates is demonstrated in Figure 5.

SUMMARY

Normal-phase adsorption chromatography has been shown to be a valuable technique for the separation of nucleobases and nucleosides, providing the mobile phase is rather polar and contains acids and/or salts. The order of elution is approximately the inverse of that usually found with reversed-phase systems. However, the separation on unmodified silica gel can be realized under isocratic conditions. This is in contrast to reversed-phase systems, where a gradient is necessary.

With respect to selectivity, unmodified silica gel is as least as good as, and for some pairs of solutes even better than, reversed-phase methods. However, the impossibility to inject aqueous samples directly into the mobile phase is a serious disadvantage which prohibits the application of normal-phase adsorption systems in certain areas of research.

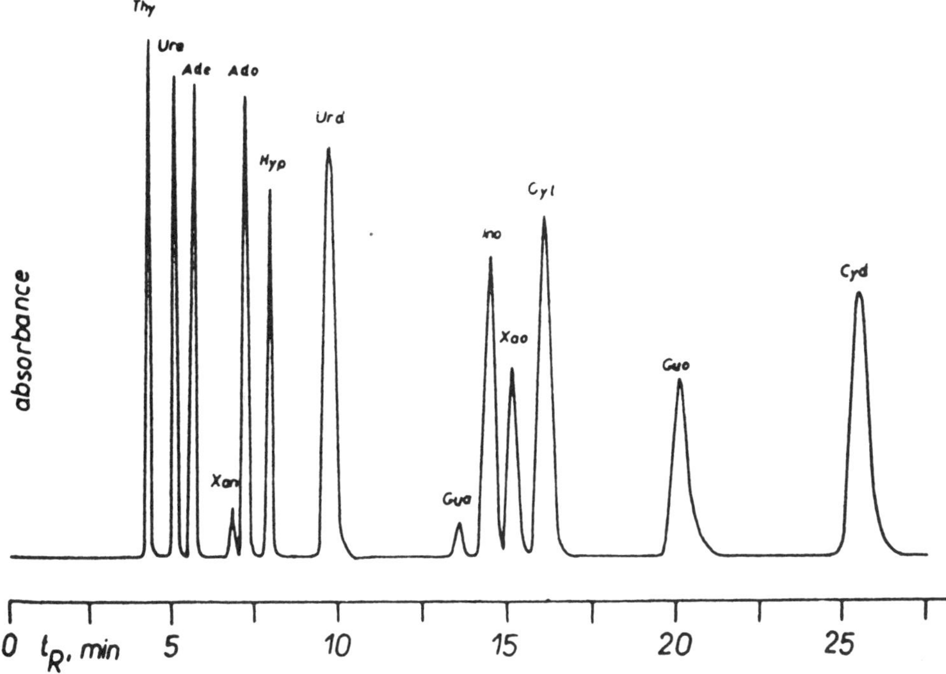

FIGURE 3. Separation of bases and nucleosides on silica gel. Stationary phase: Lichrosorb® Si 100 (250 × 4.6 mm ID); mobile phase: dichloromethane-methanol-ammoniumformate (80:18:2), pH 2.5. (From Ryba, M. and Beránek, J., *J. Chromatogr.*, 211, 337, 1981. With permission.)

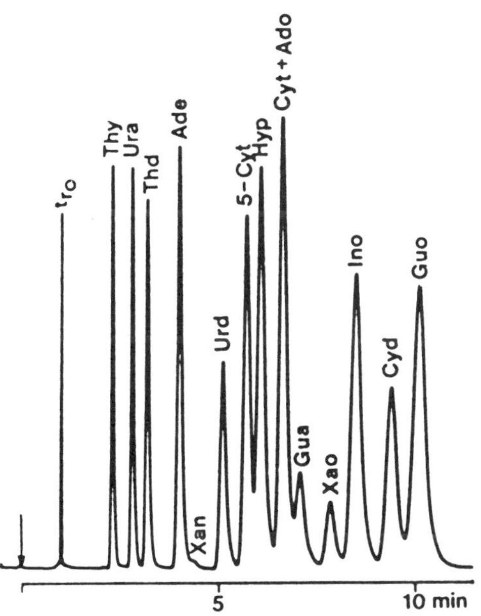

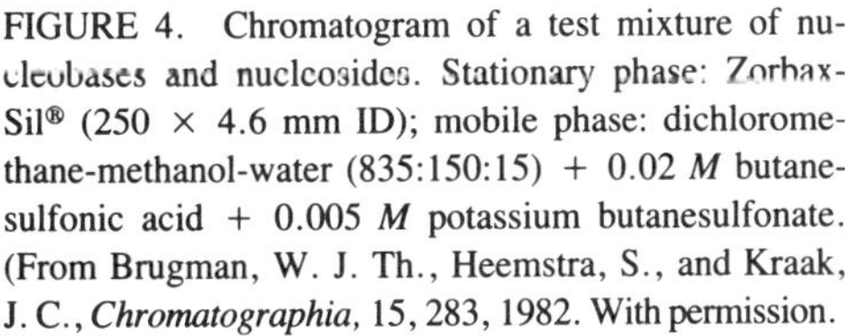
FIGURE 4. Chromatogram of a test mixture of nucleobases and nucleosides. Stationary phase: Zorbax-Sil® (250 × 4.6 mm ID); mobile phase: dichloromethane-methanol-water (835:150:15) + 0.02 *M* butanesulfonic acid + 0.005 *M* potassium butanesulfonate. (From Brugman, W. J. Th., Heemstra, S., and Kraak, J. C., *Chromatographia,* 15, 283, 1982. With permission.

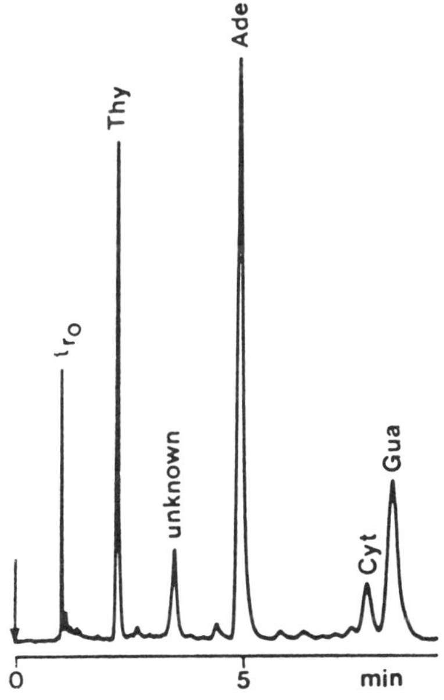

FIGURE 5. Chromatogram of calf thymus DNA hydrolysate. Stationary phase: Zorbax-Sil® (250 × 4.6 mm ID); mobile phase: dichloromethane-methanol-water (835:150:15 v/v) + 0.025 *M* butanesulfonic acid. (From Brugman, W. J. Th., Heemstra, S., and Kraak, J. C., *Chromatographia,* 15, 283, 1982. With permission.)

REFERENCES

1. **Evans, J. E., Tieckelman, H., Naylor, E. W., and Guthrie, R.,** Measurement of urinary pyrimidine bases and nucleosides by HPLC, *J. Chromatogr.*, 163, 29, 1979.
2. **Ryba, M. and Beránek, J.,** Liquid chromatographic separation of purines, pyrimidines and their nucleosides on silica gel columns, *J. Chromatogr.*, 211, 337, 1981.
3. **Brugman, W. J. Th., Heemstra, S., and Kraak, J. C.,** Optimum conditions for the separation of nucleobases and nucleosides on unmodified silica by HPLC, *Chromatographia*, 15, 283, 1982.
4. **Pfadenhauer, E. H.,** Rapid determination of some plasma oxypurines using HPLC, *J. Chromatogr.*, 81, 85, 1973.
5. **Elrassi, Z., Gonnet, C., and Rocca, J. L.,** Chromatographic studies on the influence of water and thermal treatment on the activity of silica gel, *J. Chromatogr.*, 125, 179, 1976.
6. **Brugman, W. J. Th., Heemstra, S., and Kraak, J. C.,** HPLC of organic acids on bare silica, *J. Chromatogr.*, 218, 285, 1981.
7. **Crombeen, J. P., Heemstra, S., and Kraak, J. C.,** The applicability of liquid-liquid systems in HPLC, *J. Chromatogr.*, 282, 95, 1983.
8. **Horgan, R. and Kramers, M. R.,** High performance liquid chromatography of cytokinins, *J. Chromatogr.*, 173, 263, 1979.

THE LIQUID CHROMATOGRAPHIC BEHAVIOR OF OUTER-SPHERE AND INNER-SPHERE COMPLEXES OF NUCLEOSIDES AND NUCLEOTIDES WITH METAL IONS

Aharon S. Cohen and Eli Grushka

INTRODUCTION

The interactions between metal ions and organic or inorganic ligands have been utilized for the achievement of chromatographic separations and purifications. For example, ion exchange[1-3] and ligand exchange chromatography[4-6] were able to solve some of the outstanding problems which could not be done with the more conventional forms of chromatography. In fact, reversed-phase chromatography has been frequently modified so that advantage can be taken of such interactions (viz. 7).

The nature of the interactions between any ligand and the metal cations can vary according to the chemical environment. For the purpose of the present discussion, a distinction will be made between inner-sphere and outer-sphere complexes. Inner-sphere complexes are those in which the ligands occupy the first coordination shell. It is represented schematically by the equation:

$$ML_{n-1} + L \rightarrow [ML_n] \qquad (1)$$

M and L denote metal cation and ligand, respectively. Inner-sphere complexes can cover a wide range of stability and inertness, factors which are essential for successful chromatographic separations. Outer-sphere complexes are those in which the first coordination shell is filled with one (usually) type of ligand, and a second type of ligand enters the next-nearest coordination sphere. This is shown schematically in the following equation:

$$ML_n + L' \rightarrow [ML_nL'] \qquad (2)$$

L′ represents the ligand in the outer sphere. This type of complex is frequently formed with coordinatively saturated inert complexes of Co(III) and Cr(III). At present, there is sufficient evidence that all the cationic complexes of these metals can form outer-sphere complexes with anions[8] or with molecules having donating properties.[9] For example, Co(III) inner-sphere complexes of ammonia or ethylenediamine[9,10] can form outer-sphere complexes with anionic species like PO_4^{3-} and HPO_4^{2-}.

For chromatographic purposes, advantage can be taken either of the properties of the complexes as a whole, or of their ligand exchange capabilities. This, coupled with the fact that an additional "fine tuning" can be made by utilizing selectively inner- and outer-sphere interactions, points to the potential benefits of adding metal cations to the chromatographic system. An example of the power and versatility of such an approach is the enantiomeric separation of Co(III) complexes.[11] This review will further demonstrate this point by describing the chromatographic behavior and separation through inner- and outer-sphere interactions of nucleotides and nucleosides.

The review will be divided into two parts. First, it will be shown that chromatography allows the studying of the physicochemical properties of the complexes of these compounds with metal cations. Then, it will be demonstrated that the use of metal cations offers unique possibilities for the selective separations of nucleosides and nucleotides.

Before proceeding with the topic at hand, a brief discussion on the relevant properties of nucleosides and nucleotides is in order. These compounds exhibit a large range of behaviors which are essential for their proper biofunctioning. Substituents on the purine or pyrimidine base have a substantial effect on the electronic structure of the nucleosides and nucleotides. This is evident from the spectral behavior of the $\pi \rightarrow \pi^*$ and $n \rightarrow \pi^*$ transitions.[12,13] The absorption spectrum, in the range of 190 to 300 nm, is very dependent on the pH of the solution,[14] since protonation of the base nitrogen has a strong influence on the $n \rightarrow \pi^*$ transition. The polarity of the solvent can also affect the electronic transitions, depending on the substituent and its position. The pH controls the charge of the phosphate group(s), and at pH >3 the nucleotides are ionized.

Another important property of (at least some) nucleosides and nucleotides is their ability to self-associate, or to "stack"[15-17] in aqueous solutions. By stacking it is meant the vertical aggregation, in a "head-over-tail" fashion, of these compounds in solution. Brown and Grushka[18] investigated the relationships between the chromatographic retention on reversed-phase columns and stacking of nucleotides. They found that factors affecting stacking affect retention as well.

All these factors are important in the interactions between nucleotides (and nucleosides) and metal cations. They control not only the formation of inner- and outer-sphere complexes, but also their physicochemical behavior, including chromatographic retention. The presence of metal cations can be essential to the proper functioning of nucleotides and related compounds; e.g., the adenilate enzymatic system.[19-22] Due to their importance, the interplay between metal ions and nucleotides is currently receiving a great deal of attention. Among others, Sigel and co-workers have studied extensively these interactions (viz.[22,23]). They found, for example, that metal ions can promote stacking via charge neutralization of the phosphate moiety. Chromatography is very sensitive to changes in the charge of the solute molecules, and it is not surprising, therefore, that the presence of metal ions will influence the retention of the nucleotides. The present review deals with this influence and its implications.

PHYSICOCHEMICAL PROPERTIES OF METALLO-NUCLEOTIDE COMPLEXES

Chromatography, in addition to being an excellent separation tool, can be utilized for obtaining information pertaining to the nature of the solute. Among the information which can be obtained is the formation constant of a complex which is formed in the chromatographic system. If an injected solute can form a complex with metal ions, present either in the stationary or the mobile phase, the formation constant of that complex can be calculated from the chromatographic retention and its dependence on the metal ion concentration. The chromatographic capacity ratio, k′, can be related to the complex formation constant in the following manner.[24,25] Assuming that the metal cations do not partition into the mobile phase, and that the complex formation takes place only in the mobile phase, the capacity ratio of the solute is given by:

$$k' = \frac{[Nu]_S + [NuM]_S}{[Nu]_M + [NuM]_M} \tag{3}$$

In the above equation Nu represents the nucleotide, NuM the complex, and subscript S and M indicate relevent quantities in the stationary and mobile phase, respectively. Dividing top and bottom of Equation 3 by $[Nu]_M$ the following expression is obtained:

$$k' = \frac{k_o' + k_c' K_f [M]_M}{1 + K_f [M]_M} \tag{4}$$

Table 1
A COMPARISON BETWEEN THE FORMATION CONSTANTS OF NUCLEOTIDE-METAL ION COMPLEXES OBTAINED BY CHROMATOGRAPHIC METHOD AND LITERATURE VALUES

Complex	K_f by chromatography	Literature values[a]
Part A: Data Based on the Work of Horvath et al.[24]		
ATP-Mg	2.298	2.213 (46)
GTP-Mg	2.444	2.269 (46)
ATP-Zn	3.006	3.003 (46)
ADP-Zn	2.561	2.517 (46)
GTP-Zn	2.773	—
CTP-Zn	3.519	—
CDP-Zn	3.117	—
Part B: Data Based on the Work of Grushka et al.[25]		
ADP-Mg	4.32	3.17 (46)
ATP-Mg	4.57	4.06 (46), 4.24 (47)
GTP-Mg	4.55	4.02 (46), 4.13 (47)
CDP-Mg	4.02	3.22 (46)
CTP-Mg	4.39	4.01 (46), 4.08 (47)
UTP-Mg	4.40	4.02 (46), 4.00 (47)

[a] The numbers in parenthesis are the literature references.

[M] is the concentration of the metal cation, K_f is the formation constant of the complex, and k'_o is the capacity ratio of the neat nucleotide defined as:

$$k'_o = \frac{[Nu]_S}{[Nu]_M} \tag{5}$$

k'_c is the capacity ratio of the complexed form given by:

$$k'_c = \frac{[NuM]_S}{[NuM]_M} \tag{6}$$

Equation 4 allows the calculation of the formation constant of the nucleotide-metal ion complex, provided that the concentration of the solute is no larger, in magnitude, than K_f. A complete discussion on the use of Equation 4 can be found in the work of Horváth and co-workers.[24]

Table 1 shows the K_f values of some nucleotide-cation complexes obtained by the chromatographic method. Also given in the table are literature values, obtained using other techniques. The agreement between the chromatographic results and other methods is very good.

The chromatographic method of obtaining the formation constant has several advantages. First, being a separation technique, it does not require that the solutes be absolutely pure. Thus, sample pretreatment can be kept at a minimum. Second, the data are obtained dy-

Table 2
COMPARISON OF THE BEHAVIOR OF BASE ACIDITY, AS MEASURED FROM pK_a, AND THE RETENTION, AS INDICATED BY %Δk′, FOR GTP AND UTP

Solute	pK_a[a]	%Δk′[b]
Mg (GTP)	0.2	−89
Ni (GTP)	1.15	−76
Zn (GTP)	1.40	−62
Cu (GTP)	1.9	+74
Mg (UTP)	0.2	−94
Ni (UTP)	0.6	−69
Zn (UTP)	0.99	−74
Cu (UTP)	1.7	+53

[a] Data taken from Sigel.[45]
[b] $\%\Delta k' = (k' - k'_o)/k'_o$, where k'_o is the capacity ratio in the absence of metal cation and k′ is obtained at 2 m*M* cation concentration.

Reproduced from Cohen, A. S. and Grushka, E., *J. Chromatogr.*, 318, 221, 1985. With permission.

namically, in a flowing system. Therefore, the experiment time is relatively short. Finally, the value of the formation constant is calculated from several different experimental runs. The accuracy of the method should consequently improve. There are, of course, some limitations to the technique. The biggest disadvantage is due to the fact that complexes with very large K_f values may be difficult to determine because of the requirement of low solute concentration.[24]

Chromatographic retention data can be used to characterize other properties of nucleotide-cation complexes. Sigel and workers[22,23] have devoted a great deal of time to the elucidation of nucleotide-metal ion interactions using nonchromatographic techniques. They dealt with such properties as base acidity changes due to complexation with cations, dephosphorylation rates, and their dependence on type of cation, inter- and intramolecular chelation formation with cations, etc. Cohen and Grushka[26] showed that a correlation exists between retention data and the above-mentioned parameters. For example, the elution order of any nucleotide in the presence of Mg(II) or Ni(II) or Zn(II) or Cu(II) is identical to the order of dephosphorylation rates of that nucleotide in the presence of the same cations. Another example is given in Table 2, where changes in the retention of the some nucleotides in the presence of different cations show the same trend as base acidity behavior. Similarly, it was found that general trends in the retention behavior can be indicative of intramolecular chelate formation. In that regard, it is interesting to note the extreme difference in the affect of Mg(II) ions vs. Cu(II) ions. Mg(II) ions, which are coordinated only by the phosphates, caused a decrease in retention, while Cu(II), which can promote macrochelation, increased drastically the elution times.

Clarke et al.,[27] in a somewhat different approach, showed that reversed-phase chromatography can be employed to study metallo nucleotide hydrolysis. They found that the formation of (Guo) $(NH_3)_5Ru(II)$ provides clear evidence that the metal catalyzes the cleavage of the sugar-purine bond. In their investigation the nucleotide metal complex was prepared prior to the injection on the column, and neither the stationary nor the mobile phase contained any metal additives. Thus, while not within the scope of the present review, it is mentioned here for the sake of completeness.

As indicated above, this approach to characterization of nucleotide-metal complexes is relatively new. It is sufficiently promising, however, to warrant a wider scope of investigation, particularly since it can be applied to the biologically important ternary complexes of nucleotide-metallic cation-amino acid. Research in this direction is currently being pursued, and initial results were recently reported.[26] The great advantage of the chromatographic approach lies in the fact that the required concentrations of all species are small, thus better simulating naturally occurring systems. The ease of collecting the data should also be mentioned. On the other hand, chromatography, unlike techniques such as NMR, has great difficulties in probing specific sites in the molecules.

APPLICATIONS OF INNER- AND OUTER-SPHERE COMPLEXATION FOR SEPARATION OF NUCLEOSIDES AND NUCLEOTIDES

Metal cations can be utilized for the achievement of chromatographic separations via two different approaches. In one the metal cation is bonded to the stationary phase, while in the other the cation is a component of the mobile phase. The interactions with the solutes can occur through the inner coordination sphere or the outer one. The former constitutes the technique known as ligand exchange chromatography. While conventionally performed with columns where the metal cation is a part of the stationary phase, we shall consider for the sake of this review all inner-sphere aided separations as a ligand exchange system. Outer-sphere aided separations are those in which the primary coordination shell of the metal ions is occupied by ligands other than the solutes in question, and the interactions with the solutes occur in the outer coordination shell. Again, the fully coordinated metal ion can be made available either in the mobile or in the stationary phase. This review will discuss almost exclusively the situation where the complexes are formed *in situ* in the column during the chromatographic process. Cases where the complexes are formed prior to the injection fall into an entirely different class of separations which is outside the scope of the present work. In general, it can be said that such complexes, being very stable and inert, travel down the chromatographic column in the form in which they were injected.

The exchange rate of the ligands in kinetically inert complexes, such as complexes of Co(III) and Cr(III), is very small, at least in terms of chromatographic retention times. The use of these complexes either in the stationary or mobile phase can facilitate separations via outer-sphere interactions, which can be quite labile. For that reason Chow and Grushka[28] prepared a bonded phase which contained in the terminal group a fully coordinated Co(III)-ethylenediamine complex, $Co(en)_3^{3+}$. This cobalt complex was chosen because of its well-known ability to form outer-sphere complexes with negatively charged ions.[29] The following scheme shows the synthetic route. First, silica gel (symbolized by R in the following expressions) is refluxed in dry benzene with a diamine silane reagent. This results in a diamine bonded phase:

$$\text{R–Si–OH} + (CH_3O)_3Si\,(CH_2)_3NH(CH_2)_2NH_2 \rightarrow \text{R–Si–O–Si}\,(CH_2)_3NH(CH_2)_2NH_2 \quad (7)$$

Next, cobalt dichlorodiethylenediamine is added to the freshly prepared bonded silica:

$$\text{R–Si–O–Si}\,(CH_2)_3NH(CH_2)_2NH_2 + Co(en)_2Cl_2^+ \rightarrow$$

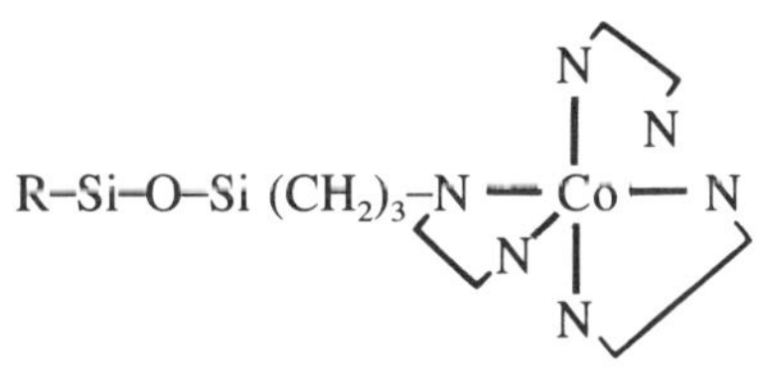

$$\begin{array}{c} \text{N} \quad \\ | \;\; \text{N} \\ \text{R–Si–O–Si}\,(CH_2)_3\text{–N}\text{—Co—N} \\ \text{N} \;| \quad \\ \text{N} \quad \end{array} \quad (8)$$

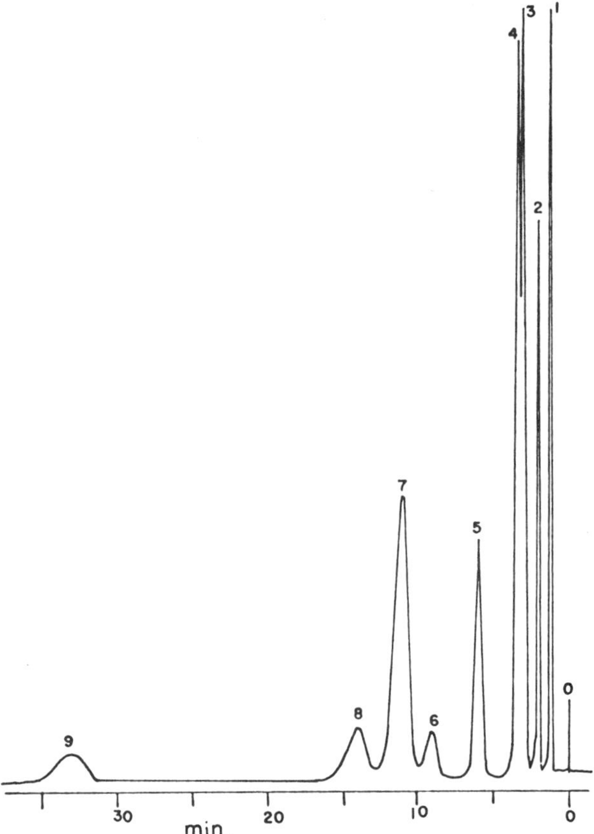

FIGURE 1. Separation of some nucleotides on the $Co(en)^{3+}$ column. Mobile phase: 0.037 *M* $Na_2HPO_4 \cdot 7H_2O$; pH 6.4, flow rate 1.0 mℓ/min. 1 = Uridine; 2 = UMP; 3 = AMP; 4 = GMP; 5 = UDP; 6 = CDP; 7 = ADP; 8 = GDP; 9 = UTP. (Reproduced from Chow, F. K. and Grushka, E., *J. Chromatogr.*, 185, 361, 1979. With permission.)

The above stationary phase was made with the aim of separating nucleotides which, because of the negatively charged phosphate group(s), could interact with the outer-coordination sphere of the cobalt complex. A typical chromatogram of some nucleotides on the cobalt column is shown in Figure 1. With this column nucleosides elute very fast. In general, the elution order of the nucleotides is mono- < di- < triphosphate, as expected from the number of negative charges of each species. Also, the pyrimidine nucleotides elute before the purine nucleotides. Hence, it is seen that the column behaves as an ion exchanger.

The long retention times of the triphosphate nucleotides are exemplified in the over 30-min elution of UTP. The long retention is in spite of the fact that the interactions between the nucleotides and the cobalt-bonded complex were done via the outer-coordination sphere.

There are several ways to shorten the analysis time. One such way is to add another complexant to the chromatographic system, which can form inner-sphere complexes with the solutes. Figure 2 shows the effect of Mg(II) ions in the mobile phase: the presence of 1 m*M* of that ion reduced the retention of the solute by a factor of 3. With Mg(II) ions GTP and ATP could also be eluted in a reasonable time. The overall retention order, however, remained as above. When Mg(II) ions are in the mobile phase, the nucleotides can form outer-sphere complexes with the cobalt compound or inner-sphere complexes with the magnesium ions. The latter seems to dominate as evident from the shorter retentions. The

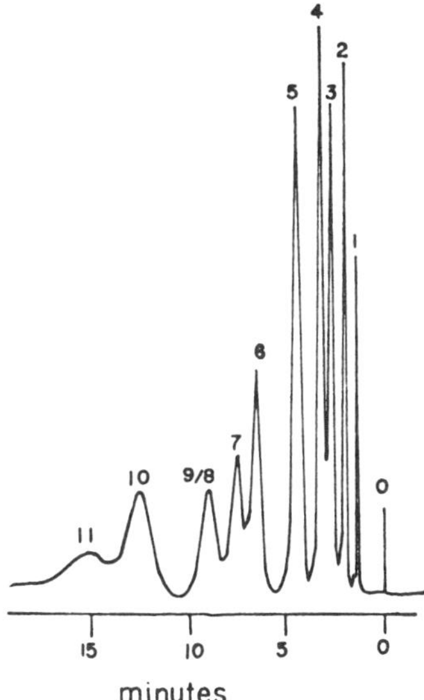

FIGURE 2. Separation of some nucleotides on the $Co(en)^{3+}$ column. Mobile phase: 0.037 *M* $Na_2HPO_4{\cdot}7H_2O$ and 0.001 *M* $MgSO_4{\cdot}7H_2O$: pH 6.4, flow rate 1.0 mℓ/min. 1 = Uridine; 2 = UMP; 3 = AMP; 4 = GMP; 5 = UDP; 6 = CDP; 7 = ADP; 8 = GDP; 9 = UTP; 10 = ATP; 11 = GTP. (Reproduced from Chow, F. K. and Grushka, E., *J. Chromatogr.*, 185, 361, 1979. With permission.)

dependence of the retention on the Mg(II) ion concentration and on the pH, as can be seen in Table 3, tends to support the above statement. The relevant equilibrium process taking place in the column can, most likely, be represented by:

$$RR\text{–}Co(en)_3\text{–}Nu + \text{anion} + Mg(II) \rightleftarrows RR\text{–}Co(en)_3\text{–anion} + NuMg \tag{9}$$

RR represents the silica gel and the organic part of the bonded phase, and Nu indicates nucleotide. The anion in the above expression is most likely HPO_4^{2-}. The data in Table 3 are in line with the fact that magnesium ions can form a strong complex with the triphosphate nucleotides and a weak one with the monophosphates. This is shown graphically in Figure 3. Plots such as in Figure 3 can be used in the optimization of nucleotide separation by constructing selectivity diagrams similar to Laub-Purnell "window diagrams".[30]

The large effect of the Mg(II) ions prompted these workers to investigate a conventional C18 column instead of the $Co(en)_3$ column.[28] Reversed-phase columns are routinely used for the analysis of nucleotides, and their properties with regard to pH and ionic strength changes are well understood. On the other hand, the addition of magnesium ions for controlling and affecting the separation is not a common occurrence. If the cobalt moiety is removed, the outer-sphere complexes of the nucleotides do not exist and the selectivity will be manipulated by inner-sphere interactions with the Mg(II) ions. These can occur either in the stationary phase or in the mobile phase.

The significant effect of the Mg(II) ions in the mobile phase is shown in Table 4. It was found that, as with the cobalt phase, the magnesium ions decreased the retention of the nucleotides. The decrease was the most pronounced in the case of the triphosphate nucleo-

Table 3
EFFECT OF THE AMOUNT OF $MgSO_4{\cdot}7H_2O$ IN THE MOBILE PHASE ON THE k′ VALUES OF NUCLEOTIDES AT pH VALUES OF 6.4 AND 5.4 IN THE $Co(en)^{3+}$ SYSTEM

	2.03 m*M* Mg(II)		1.2 m*M* mG(II)		1.00 m*M* Mg(II)		No Mg(II)	
Solute	pH = 6.4	pH = 5.4	pH = 6.4	pH = 5.4	pH = 6.4	pH = 5.4	pH = 6.4	pH = 5.4
AMP	0.93	1.27	1.05	1.40	1.29	1.50	1.28	1.67
UMP	0.37	0.57	0.54	0.60	0.58	0.76	0.75	0.75
CMP	0.80	1.14	0.94	1.17	0.94	1.32	1.08	1.50
GMP	1.11	1.36	1.28	1.50	1.47	1.74	1.69	1.78
ADP	2.75	8.71	3.30	10.20	4.72	11.5	7.42	11.7
UDP	1.35	4.11	1.75	4.64	2.44	5.61	3.76	5.70
CDP	2.36	6.50	3.00	7.93	3.45	10.5	6.04	10.8
GDP	3.23	10.6	4.21	12.4	5.82	14.4	9.98	15.1
UTP	2.34	17.3	3.09	22.3	5.62	28.3	24.9	64.8
CTP	3.59	26.4	4.88	[a]	8.12	[a]	47.3	[a]
ATP	4.03	[a]	5.24	[a]	8.48	[a]	[a]	[a]
GTP	4.93	[a]	6.50	[a]	10.6	[a]	[a]	[a]

[a] Not eluted within reasonable time (k′ values greater than 50).

Reproduced from Chow, F. K. and Grushka, E., *J. Chromatogr.*, 185, 361, 1979. With permission.

tides, and the least with the monophosphates. The roughly equivalent relative change in the k′ values of the pyrimidine nucleotides and the purine nucleotides points out the importance of the phosphate groups in determining the retention behavior. The importance of the pH should also be noticed. Because of the varied ionization properties of the nucleotides, it is not surprising that there is not a clear-cut pattern to the effect of the pH.

Figures 4 and 5 show another characteristic of the magnesium additive. On the whole, the efficiencies of the Mg(II)-containing column were superior, at times quite drastically so, to the conventional reversed-phase column. Advantage can be taken of the unequal effect of Mg(II) ions on all the solutes to "fine tune" a particular separation by the correct choice of the magnesium concentration.

The above points were further demonstrated by Grushka and Chow[25] with the aid of especially synthesized dithiocarbamate bonded phase. The stationary phase was prepared as follows: silica gel and trimethoxypropylamine silane were refluxed in dry benzene to give:

$$\text{R–SiOH} + (CH_3O)_3\text{Si}\ (CH_2)_3NH_2 \rightarrow \text{R–Si–O–Si}\ (CH_2)_3NH_2 \quad (10)$$

Next, CS_2 is reacted with the amine bonded phase in an alkaline environment:

$$\text{R–Si–O–Si}\ (CH_2)_3NH_2 + CS_2 \rightarrow \text{R–Si–O–Si}\ (CH_2)_3NHCS_2^- \quad (11)$$

In this column Mg(II) ions can be coordinated by the carbamate stationary phase. Yet, Figure 6 shows that the effect of the metal additive in this column is similar to that in the reversed-phase system. Again, the change is largest with the triphosphate nucleotides. Utilizing pH and Mg(II)-induced changes in the retention, the separations of complicated mixtures of nucleosides and nucleotides could be achieved with relative ease using simple pH step-gradients, e.g., the chromatogram in Figure 7.

A different approach was employed by Karger and co-workers.[31,32] They added 4-dodecyldiethylenetriamine (symbolized by C_{12}-dien from now on) and either Zn(II) or Cd(II) ions

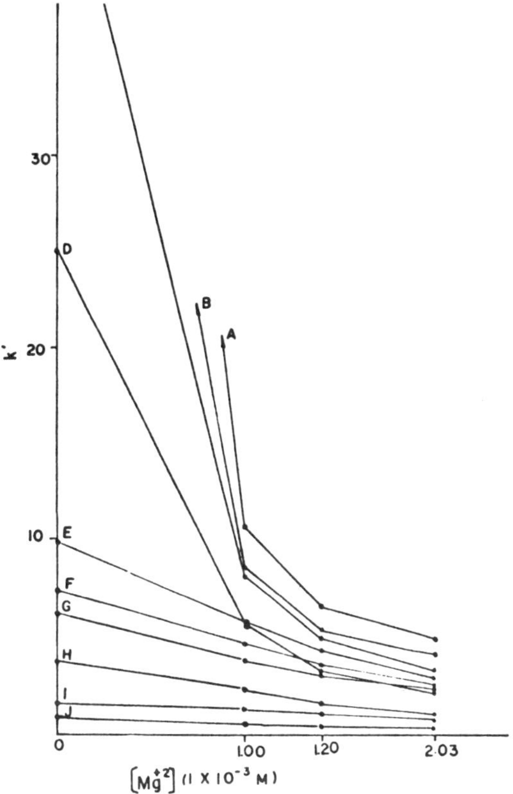

FIGURE 3. The effect of the concentration of Mg(II) on the k' values of some nucleotides on the $Co(en)^{3+}$ column. Mobile phase: 0.037 *M* $Na_2HPO_4 \cdot 7H_2O$; pH 6.4. A = GTP; B = ATP; C = CTP; D = UTP; E = GDP; F = ADP; G = CDP; H = UDP; I = GMP; J = UMP. (Reproduced from Chow, F. K. and Grushka, E., *J. Chromatogr.*, 185, 361, 1979. With permission.)

to an acetonitrile-water mobile phase of a C18 reversed-phase system. The cation is complexed by the C_{12}-dien as shown in Figure 8. The rationale behind this reagent is as follows: (1) the complex formed is charged which may aid in the selectivity; (2) in the absence of a solute, a modifier molecule occupies the fourth coordination position; (3) solutes such as nucleotides will replace the modifier molecule in the complex, thus, forming an inner-sphere complex; (4) Zn(II) and Cd(II) were chosen as the metal ions, since they allow relatively (to cupric and nickel ions) easy exchange of ligands, hence, improving the chromatographic efficiency of the system. Many properties of the exchange site can be altered by the choice of the metal cation. Among these are (1) the charge of the complex, (2) the size and conformation of the chelate ring, and (3) the structure of the inner coordination sphere; e.g., a change from a tetra coordination to hexa-coordinated center. Of course, the choice of metal cations in liquid chromatography is limited by their UV absorbance, since they can interfere with detection.

Table 5 shows some chromatographic properties of the C_{12}-dien-Zn(II) and the C_{11}-dien-Cd(II) systems in the separation of some nucleotides. The experimental conditions were identical in both cases with the exception of the metal cation. It is seen that the cyclic nucleotide elutes much before the other nucleotides. This is the result of the lesser charge on the former solute. As expected, the retention is weaker in the cadmium system. Similarly,

Table 4
EFFECT OF pH VALUES ON k′ VALUES IN THE C_{18} SYSTEM WITH MOBILE PHASES OF 0.1 *M* KH_2PO_4 WITH EITHER 0.017 *M* $MgSO_4{\cdot}7H_2O$ OR 0.017 *M* Na_2SO_4

	pH = 4.5		pH = 5.0		pH = 5.5		pH = 6.0		pH = 6.45	
Solute	Mg(II)	Na(I)	Mg(II)	Na(I)	Mg(II)	Na(I)	Mg(II)	Na(I)	Mg(II)	Na(I)
UMP	0.96	1.03	0.68	0.94	0.80	0.81	0.72	0.82	0.69	0.82
CMP	0.71	0.71	0.50	0.61	0.50	0.42	0.42	0.52	0.42	0.49
GMP	2.38	2.37	2.01	2.35	2.16	2.51	1.63	2.69	2.04	2.29
AMP	5.90	5.92	5.88	6.50	5.88	6.10	5.19	6.12	5.24	5.08
UDP	0.54	0.39	0.42	0.48	0.31	0.45	0.30	0.40	0.38	0.46
CDP	0.36	0.39	0.21	0.30	0.23	0.28	0.19	0.27	0.25	0.30
GDP	0.95	2.46	1.61	2.46	1.90	2.72	1.16	2.66	1.20	2.08
ADP	1.79	3.85	3.19	3.89	3.12	3.82	3.10	3.78	3.20	—
UTP	0.75	1.34	0.32	0.76	0.35	0.44	0.37	—	0.45	0.41
CTP	0.54	0.93	0.25	0.52	0.23	0.40	0.21	0.36	0.22	0.29
ATP	5.25	—	3.21	6.17	3.25	5.20	3.10	4.30	3.49	3.18
GTP	3.2	[a]	1.66	[a]	1.69	[a]	2.40	5.96	1.03	4.12
Uridine	3.92	3.82	3.30	3.74	3.75	3.56	4.00	3.66	3.72	3.60
Cytidine	2.33	2.37	2.08	2.24	2.34	2.19	2.50	2.58	2.40	2.31
Adenosine	[a]	[a]	[a]	[a]	[a]	[a]	43.3	[a]	[a]	[a]
Guanosine	14.1	12.9	11.6	13.0	14.7	15.5	12.7	12.9	14.3	11.6

[a] Not eluted within reasonable time.

Reproduced from Chow, F. K. and Grushka, E., *J. Chromatogr.*, 185, 361, 1979. With permission.

the chromatographic efficiencies and peak symmetry were superior with the Cd(II) system, a fact pointing out to the more labile nature of the nucleotide C_{12}-dien-Cd(II) complex. Figure 9 shows a typical separation which can be achieved with that reagent.

Corradini et al.[33] investigated a chromatographic system similar to that of Chow and Grushka.[28] They used a cross-linked dextran gel (Sephadex G-75) ion-exchange material loaded with an amino cobalt complex, for the outer-sphere separations of nucleotides. They reported the separation of CMP, GMP, and AMP, as well as of two pairs of dinucleotides; C_PU-A_PU and U_PA-A_PU. It was found that on the amino-polysaccharide column without the cobalt, these solutes were strongly retained. The cobalt complex in the form of $Co(tetren)_3^{3+}$, is essential for reasonable retention times. The interactions taking place in this column were assumed to be outer-sphere complexation between the solutes and the cobalt species.

Hubert and Porath[34,35] took advantage of inner-sphere complexation for the separations of nucleotides and oligonucleotides using copper(II) chelated gels. Their results seem to indicate that the interactions with the cupric stationary phase are via the purine and pyrimidine bases of the solutes. The phosphate part did not seem to control the retention process. The general retention order in this system was bases > nucleosides > nucleotides. The authors attributed that order to the importance of base-Cu(II) interactions. Figures 10 and 11 demonstrate the separation of CpA + ApC and of CpG + GpC on the copper column. Although the bases play a major role in determining the retention, the conformation of the molecule should not be neglected. ApC and CpA models, as well as theoretical calculations,[36] indicate very small differences between these two molecules. Yet, the good resolutions shown in Figures 10 and 11 point to a different environment of the binding site, which is probably

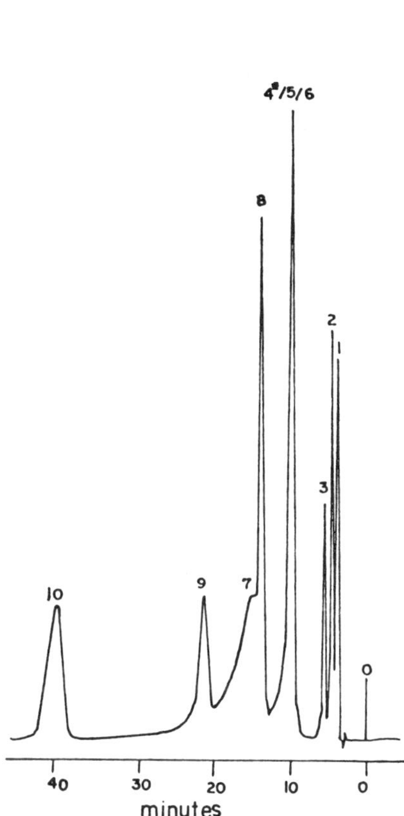

FIGURE 4. Separation of some nucleosides and nucleotides on the C_{18} column. Mobile phase: 0.1 *M* KH_2PO_4 with 0.017 *M* Na_2SO_4; pH 6.0, flow rate 1.1 mℓ/min. 1 = CDP; 2 = CMP; 3 = UMP; 4 = GDP; 5 = GMP; 6 = cytidine; 7 = ATP; 8 = uridine; 9 = AMP; 10 = guanosine. (Reproduced from Chow, F. K. and Grushka, E., *J. Chromatogr.*, 185, 361, 1979. With permission.)

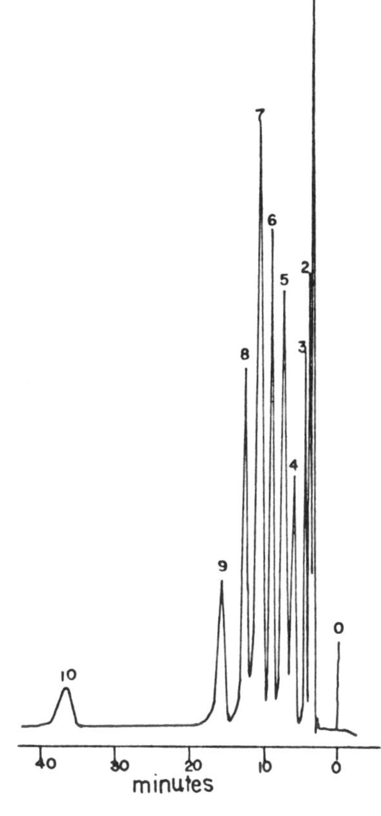

FIGURE 5. Separation of some nucleosides and nucleotides on the C_{18} column. Mobile phase: 0.1 *M* KH_2PO_4 with 0.017 *M* $MgSO_4{\cdot}7H_2O$; pH 6.0, flow rate 1.0 mℓ/min. 1 = CDP; 2 = CMP; 3 = UMP; 4 = GDP; 5 = GMP; 6 = cytidine; 7 = ATP; 8 = uridine; 9 = AMP; 10 = guanosine. (Reproduced from Chow, F. K. and Grushka, E., *J. Chromatogr.*, 185, 361, 1979. With permission.)

at the N7 position of the purine ring.[37] Thus, the complexes formed have distinguishable conformations. Undoubtedly, however, hydrophobic interactions influence the retention as well, and the order may be controlled to a large extent by the polarity of the solutes.

Inner-sphere interactions were also used by Bij and Lederer[38] for the separations of nucleotides and their bases on cellulose layers. The R_f values of the nucleotides only were affected by the presence of divalent metal ions in the cellulose. The increase in the R_f values was in the order Zn(II) > Mn(II) > Mg(II).

Goldstein and co-workers[4,5] used inner-sphere coordinations for the analysis and separations of nucleic acid constituents. They found that nucleotides were not retained on Cu(II) chelated resin and, therefore, they could easily separate them from nucleosides and nucleic bases. Nucleosides of low basicity were eluted from the Cu(II)-loaded resin with water as eluent. To elute nucleosides of higher basicity 1. 0 *M* NH_4OH solution had to be used as eluent. For the bases the mobile phase was 2.4 *M* ammonia solution.

White and Zenser[39] examined the retention of nucleotides on an alumina column at pH 7.6. Generally, they found that all nucleotides, except the 3′,5′-monophosphates, were

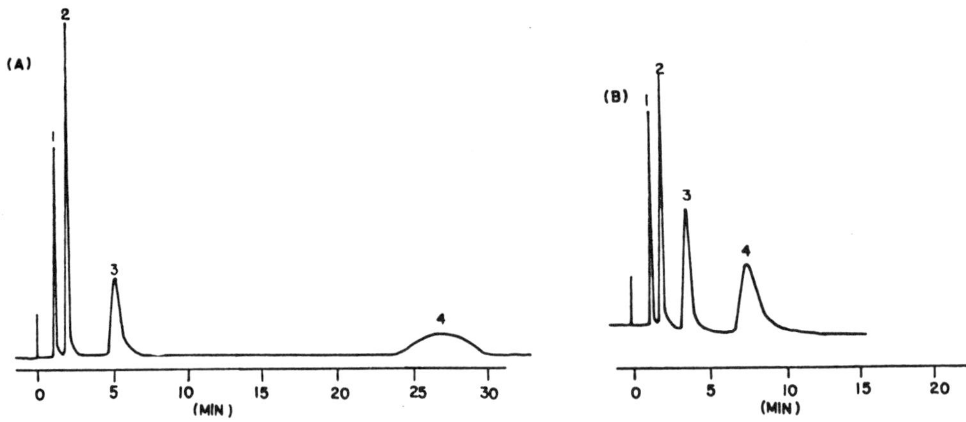

FIGURE 6. Separation of some nucleosides and nucleotides. (A) Mobile phase: 0.137 *M* KH_2PO_4 with no Mg(II), pH 60, flow rate 1.5 mℓ/min; (B) mobile phase: 0.137 *M* KH_2PO_4 with 8.1 × 10^{-4} *M* $MgSO_4{\cdot}7H_2O$, pH 6.0, flow rate 1.5 mℓ/min. Peaks: 1 = uridine; 2 = UMP; 3 = UDP; 4 = UTP. (Reproduced from Grushka, E. and Chow, F. K., *J. Chromatogr.*, 199, 283, 1980. With permission.)

retained according to their anionic character at that pH. They also found that their system could discriminate between univalent and divalent nucleotides, the latter being more strongly adsorbed on the alumina then the former.

Recently, Hollis et al.[40] used Mg(II) inner-sphere complexation for the separation of nucleotides, nucleosides, and their bases. They employed 4 m*M* Mg(II) ions in a tris buffer mobile phase on a Zorbax® ODS column. Some of their results are given in Figure 12, where the separation of AMP, ADP, and ATP from different sources is shown. This separation, they maintain, was impossible to obtain in the absence of Mg(II)/tris from the mobile phase.

SUMMARY AND CONCLUSIONS

Metal ion additives, in many forms, have been used successfully in liquid chromatography for the achievement of improved separations. The concept of imposing secondary chemical equilibria on the (sometimes) primary distribution process has been proven to be a powerful approach to the manipulation and control of difficult separations. Ligand exchange chromatography, which is based on this superposition, has advanced greatly since its inception by Helfferich[41] in 1961 and, as was seen here, has developed both in scope and in nuances. Many significant separations have been reported since then (viz. 42 to 44). Initially, the usefulness of ligand exchange chromatography, in terms of peak capacity and fast retention times, was limited by the nature of the solid support, which was made of organic resins. The introduction of silica gel impregnation improved the situation to a great extent. Difficulties still remain as far as system stability and reproducibility are concerned, mainly due to metal ion leeching. The use of metal ions as part of bonded phases, or the use of metal ion additives to the mobile phase, has all but eliminated these difficulties.

The use of outer-sphere complexes for separations is a relatively new approach to ligand exchange chromatography. The retention mechanisms involved with outer-sphere formation are being actively researched, and several key questions are still unanswered; for example, (1) the nature of the interactions between inner- and outer-sphere ligands, (2) the configuration of the outer-sphere coordination shell, and (3) any inner-sphere rearrangement during the formation of the outer-sphere complex. All these questions notwithstanding, the success of this approach lies in the enhanced separation "fine tuning" capabilities and in the ability to control the retention times. Since many biologically important molecules are known to

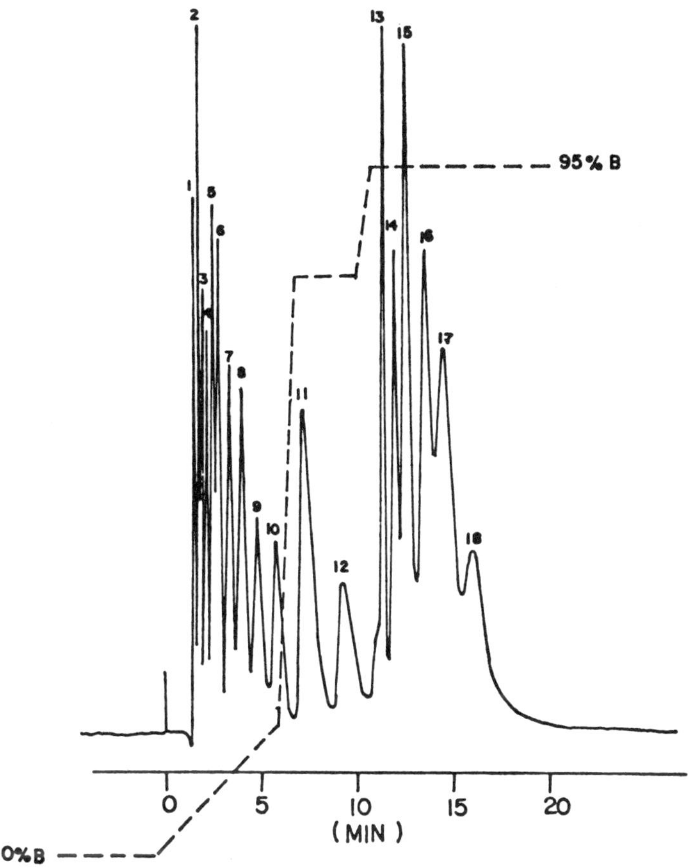

FIGURE 7. Separation of some nucleosides and nucleotides by dithiocarbamate column using gradient elution. Mobile phase A: 0.137 *M* KH_2PO_4, 0.002 *M* Mg(II) at pH 5.5. Mobile phase B: 0.137 *M* KH_2PO_4, 0.002 *M* Mg(II) at pH 7.0. Gradient profile is shown by the broken line. Peaks: 1 = uridine; 2 = thymidine; 3 = guanosine; 4 = adenosine; 5 = deoxyadenosine; 6 = UMP; 7 = CMP; 8 = dCMP; 9 = GMP; 10 = AMP; 11 = XMP; 12 = UDP; 13 = CDP; 14 = UTP; 15 = CTP; 16 = ADP; 17 = XDP; 18 = XTP. (Reproduced from Grushka, E. and Chow, F. K., *J. Chromatogr.*, 199, 283, 1980. With permission.)

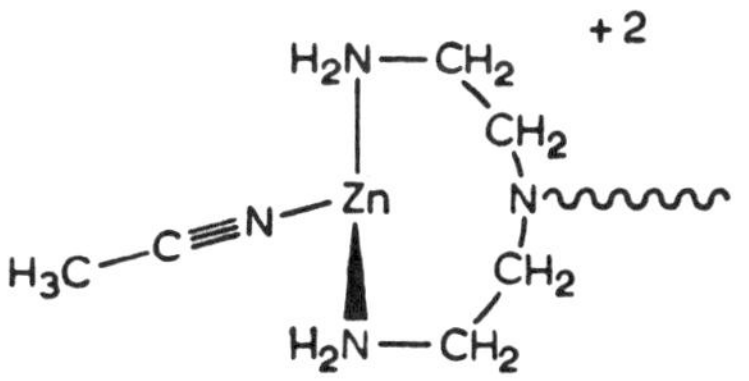

+2
$H_2N—CH_2$
CH_2
$H_3C—C{\equiv}N—Zn$ N
CH_2
$H_2N—CH_2$

FIGURE 8. Four-coordinated Zn(c_{12}-dien)$(CH_3CN)^{2+}$ complex. (Reproduced from Cooke, N. H. C., Viavattene, R. L., Eksteen, R., Wong, W. S., Davies, G., and Karger, B. L., *J. Chromatogr.*, 149, 391, 1978. With permission.)

Table 5
COMPARISON OF PERFORMANCE AND RETENTION BETWEEN C_{18}-dien-Cd(II) AND C_{18}-dien-Zn(II) WITH WEAKLY AND STRONGLY INTERACTING LIGANDS

	10^{-3} *M* C_{12}-dien-Zn(II)				10^{-3} *M* C_{12}-dien-Cd(II)			
Solutes	**k′**	**α**	**N**	**A_s**	**k′**	**α**	**N**	**A_s**
Adenosine-3′,5′-CMP	0.17		1900	1.4	0.12		1550	1.8
		12.7				2.33		
Adenosine-5′-MP	2.16		250	4.3	0.28		1250	2.0
Cytidine-2′-MP	1.82		600	1.7	0.97		—	—
		1.38				1.0		
Cytidine-3′-MP	2.51		500	3.0	0.97		—	—
Guanosine-2′-MP	4.10		700	2.5	0.97		1200	1.4
		1.34				1.58		
Guanosine-3′-MP	5.48		600	2.0	1.53		1050	1.5
Uridine-2′-MP	4.68		—	—	0.52		—	—
		1.12				1.17		
Uridine-3′-MP	5.24		—	—	0.61		—	—

Note: Common conditions are 10 μm Lichrospher® C_8 (20 cm × 4.6 mm ID), acetonitrile-water (30:70), 0.13 *M* NH_4Ac, pH 7.1, 30°C, flow rate 2 mℓ/min. A_s = asymmetry; CMP = cyclic monophosphate; MP = monophosphate.

Reproduced from Karger, B. L., Wong, W. S., Viavattene, R. L., LePage, J. N., and Davies, G., *J. Chromatogr.*, 167, 253, 1978. With permission.

form complexes with metal cations, it is surprising that not more work is described using ligand exchange. It is anticipated, however, that more advantage will be taken of the technique, in particular, since it has been demonstrated clearly now that ligand exchange chromatography, in addition to the separation, can yield important information about the properties of the complexes and, perhaps, can help in shedding some light on the biological activity of such complexes.

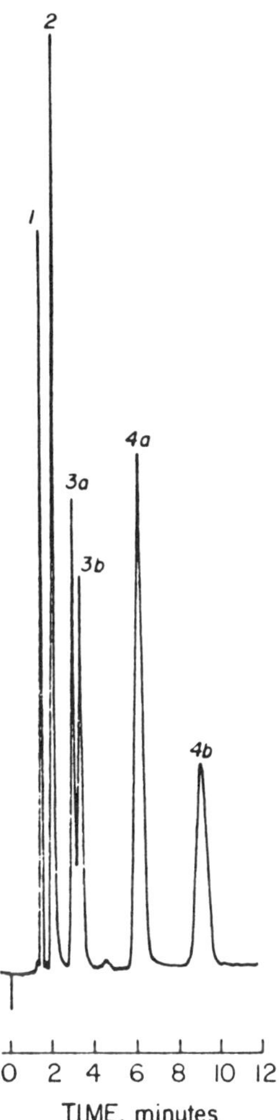

FIGURE 9. Separation of monophosphate nucleotides by C_{12}-dien-Cd(II) chromatography. Conditions: 10^{-3} *M* C_{12}-dien-Cd(II), 0.13 *M* NH_4Ac, 25/75 AN/H_2O, pH 7.1, 30°C; flow rate 2 mℓ/min. Solutes (1 to 10 μg): 1 = adenosine-3′,5′-cyclic monophosphate; 2 = adenosine-5′-monophosphate; 3a and 3b = uridine-2′ and 3′-monophosphate; 4a and 4b = guanosine-2′ and 3′-monophosphate. (Reproduced from Karger, B. L., Wong, W. S., Viavattene, R. L., LePage, J. N., and Davies, G., *J. Chromatogr.*, 167, 253, 1978. With permission.)

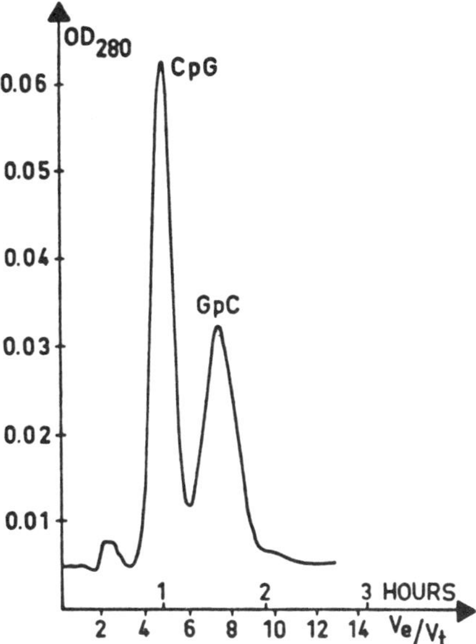

FIGURE 10. Fractionation of CpG (87.5 μg/35 μℓ + GpC (87.5 μg/35 μℓ). Copper column: V_t = 2.75 mℓ; flow rate 13.2 mℓ/hr. Buffer: 0.1 *M* ethylmorpholine-acetic acid (pH 6.5), 0.2 *M* $MgSO_4$. (Reproduced from Hubert, P. and Porath, J., *J. Chromatogr.*, 206, 164, 1981. With permission.)

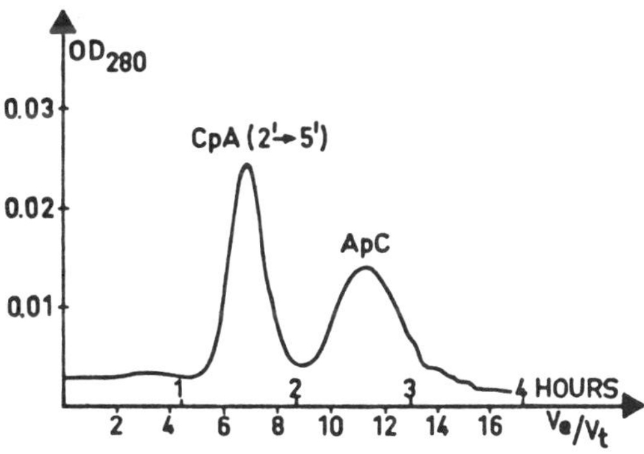

FIGURE 11. Fractionation of ApC (87.5 μg/35 μℓ) + CpA (87.5 μg/35 μℓ). Copper column: V_t = 4.5 mℓ; flow rate 19.2 mℓ/hr. Buffer: 0.1 *M* ethylmorpholine-acetic acid (pH 6.5), 0.2 *M* $MgSO_4$. (Reproduced from Hubert, P. and Porath, J., *J. Chromatogr.*, 206, 164, 1981. With permission.)

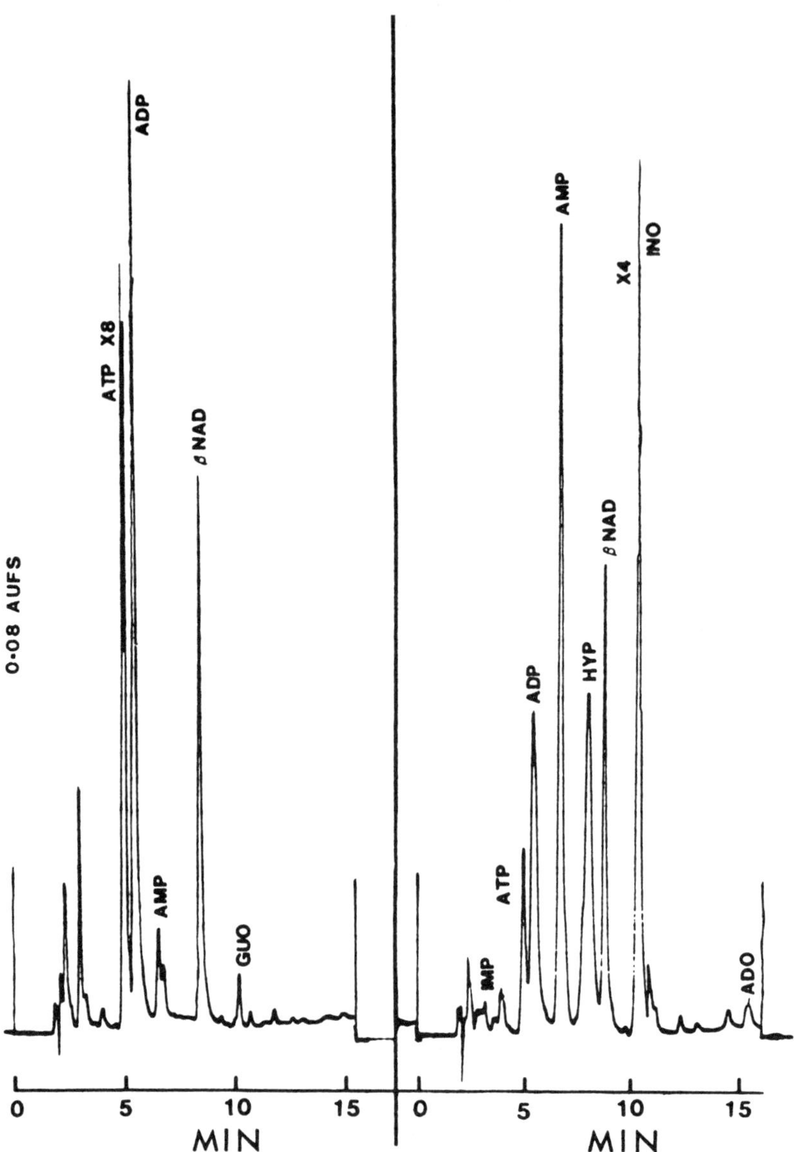

FIGURE 12. Chromatograms of adenine metabolites in extracts of normal (left) and 60-min autolyzed (right) dog myocardium, run on modified Zorbax® ODS columns. Absorbance units full scale, 0.08. ×4 and ×8 indicate peak heights four and eight times the scale indicated. (Reproduced from Holliss, D. G., Humphrey, S. M., Morrison, M. A., and Seelye, R. N., *Anal. Lett.*, 17(B18), 2047, 1984. With permission.)

REFERENCES

1. **Hartwick, R. A. and Brown, P. R.,** The performance of microparticals chemically bonded anion exchange resin in the analysis of nucleotides, *J. Chromatogr.*, 112, 651, 1975.
2. **Nurakami, F., Rokushika, S., and Hatano, H.,** Cation exchange chromatography of nucleotides, nucleosides and nucleic bases, *J. Chromatogr.*, 53, 584, 1970.
3. **Breter, H. J.,** The quatitative determination of metabolites of 6-mercaptopurine in biological materials, *Anal. Biochem.*, 80, 9, 1977.

4. **Goldstein, G.,** Ligand exchange chromatography of nucleotides, nucleosides and nucleic acid bases, *Anal. Biochem.*, 20, 477, 1967.
5. **Burtis, C. A. and Goldstein, G.,** Terminal nucleosides assay of ribonucleic acid by ligand exchange chromatography, *Anal. Biochem.*, 23, 502, 1968.
6. **Shimoura, K., Dickson, L., and Walton, H. F.,** Separation of amines by ligand exchange. IV. Ligand exchange with chelating resins and cellulosic exchangers, *Anal. Chim. Acta,* 37, 102, 1967.
7. **Fujimura, K.,** Studies in ligand exchange chromatography, *J. Chromatogr.*, 85, 101, 1973.
8. **Mason, S. F. and Norman, B. J.,** Optical rotatory power of coordination compounds. VI. The effects of ion association on the circular dichroism spectra of tris diamine complex metal ions, *J. Chem. Soc.*, A, 307, 1966.
9. **Larsson, R.,** Association of ammonia with hexamine Co(III) ion in aqueous solution, *Acta Chem. Scand.*, 12, 708, 1958.
10. **Fung, B. M.,** Proton magnetic resonance study of some Co(III) complexes with ethylenediamine. Evidence for secondary coordination shell, *J. Am. Chem. Soc.*, 89, 5788, 1967.
11. **Gaal, J. and Inczedy, J.,** Chromatographic separation of optical isomers by means of outer-sphere complex formation reaction, *Talanta,* 23, 78, 1976.
12. **Kleinwachter, V., Drubnik, J., and Augerstein, L.,** Spectroscopic studies of purines. III. Properties of purines substituted at the sixth and ninth positions, *Photochem. Photobiol.*, 6, 133, 1967.
13. **Kleinwachter, V., Drubnik, J., and Augerstein, L.,** Spectroscopic properties of the lowest-lying excited states of 2-aminopyrimidine, cytosine, uracil and their derivatives, *Photochem. Photobiol.*, 5, 579, 1966.
14. **Fox, J. J., Cavalieri, L. F., and Chang, N.,** The identification of cytidylic acids by spectrophotometric methods, *J. Am. Chem. Soc.*, 75, 4315, 1953.
15. **Bugger, C.,** *The Jerusalem Symp. on Quantum Chemistry and Biochemistry, Vol. IV,* Israel Academy of Sciences and Humanities, Jerusalem, 1972, 194.
16. **Ts'o, P. O. P., Melvin, I. S., and Olsen, A. C.,** Interaction and association of bases and nucleotides in aqueous solutions, *J. Am. Chem. Soc.*, 85, 1289, 1963.
17. **Ts'o, P. O. P., Kondo, N. S., Robins, R. K., and Broom, A. D.,** Interaction and association of bases and nucleotides in aqueous solutions. VI. Properties of 7-methylinosine as related to the nature of stacking interaction, *J. Am. Chem. Soc.*, 91, 5625, 1969.
18. **Brown, P. R. and Grushka, E.,** Structure-retention relations in the reversed phase high performance liquid chromatography of purine and pyrimidine compounds, *Anal. Chem.*, 52, 1210, 1980.
19. **Frey, C. M. and Stuehr, J. E.,** Interaction of divalent metal ions with inorganic and nucleoside phosphates. V. Kinetics of Ni(II) with $HP_3O_{10}^{4-}$, cytidine 5′-triphosphate, $HP_2O_7^{3-}$ and cytidine 5′-diphosphate, *J. Am. Chem. Soc.*, 100, 134, 1978.
20. **Eichhorn, G. L., Ed.,** *Inorganic Biochemistry, Vol. 1 and 2,* Elsevier, New York, 1973.
21. **Sigel, H., Ed.,** *Metal Ions in Biological Systems, Vol. 1 to 11,* Marcel Dekker, New York, 1973 to 1980.
22. **Scheller, K. H., Hofstetter, F., Mitchell, P. R., Prijs, B., and Sigel, H.,** Macrochelate formation in monomeric metal ion complexes of nucleotides 5′-triphosphates and the promotion of stacking by metal ions. Comparison of the self association of purine and pyrimidine 5′-triphosphates using proton NMR, *J. Am. Chem. Soc.*, 103, 247, 1981.
23. **Sigel, H., Fischer, B. E., and Farkas, E.,** Metal ion promoted hydrophobic interactions between nucleotides and amino acids mixed ligand adenosine 5′-triphosphate/metal ion (II)/L-leucinate and related ternary complexes, *Inorg. Chem.*, 22, 925, 1983.
24. **Horvath, Cs., Melander, W., and Nahum, A.,** Measurement of association constants for complexes by reversed phase high performance liquid chromatography, *J. Chromatogr.*, 186, 371, 1979.
25. **Grushka, E. and Chow, F. K.,** High performance liquid chromatography of nucleotides Mg(II) complexes. Separation and mechanism of separation, *J. Chromatogr.*, 199, 283, 1980.
26. **Cohen, A. S. and Grushka, E.,** Structure retention relationships: the chromatographic behavior and the properties of nucleotides-cation complexes, *J. Chromatogr.*, 318, 221, 1985.
27. **Clarke, M. J., Coffey, K. F., Perpall, H. J., and Lyon, J.,** Analysis of metallonucleotide hydrolysis by high performance liquid chromatography, *Anal. Biochem.*, 122, 404, 1982.
28. **Chow, F. K. and Grushka, E.,** High performance liquid chromatography of nucleotides and nucleosides using outersphere and innersphere metal solute complexes, *J. Chromatogr.*, 185, 361, 1979.
29. **Ilcheva, L., Georgieva, M., Bozadziev, P., and Karshev, I.,** Polarographic study of some outersphere complexes of trisethylenediaminocobalt(III) and hexamine cobalt(III) ions, *J. Inorg. Nucl. Chem.*, 39, 1369, 1977.
30. **Chow, F. K. and Grushka, E.,** The use of outersphere and innersphere metal solute complexes in HPLC, in *Silylated Surfaces,* Leyden, D. E. and Collins, W., Eds., Gordon and Breach, New York, 1978, 301.
31. **Cooke, N. H. C., Viavattene, R. L., Eksteen, R., Wong, W. S., Davies, G., and Karger, B. L.,** Use of metal ions for selective spearations in high performance liquid chromatography, *J. Chromatogr.*, 149, 391, 1978.

32. **Karger, B. L., Wong, W. S., Viavattene, R. L., LePage, J. N., and Davies, G.,** Reversed phase high performance liquid chromatography using metal chelate additives, *J. Chromatogr.*, 167, 253, 1978.
33. **Corradini, D., Similaldi, M., and Messina, A.,** Outersphere ligand exchange chromatography of nucleotides and related compounds on modified polysaccharide gel, *J. Chromatogr.*, 235, 273, 1982.
34. **Hubert, P. and Porath, J.,** Metal chelate affinity chromatography. I. Influence of various parameters on the retention of nucleotides and related compounds, *J. Chromatogr.*, 198, 247, 1980.
35. **Hubert, P. and Porath, J.,** Metal chelate affinity chromatography. II. Group separation of mono- and dinucleotides, *J. Chromatogr.*, 206, 164, 1981.
36. **Lee, C. H., Charney, E., and Tinoco, I.,** Conformation of dinucleotide phosphates in aqueous solution, *Biochemistry,* 18, 5636, 1979.
37. **Sigel, H., Ed.,** *Metal Ions in Biological Systems, Vol. 8,* Marcel Dekker, New York, 1979.
38. **Bij, K. E. and Lederer, M.,** Thin layer chromatography of some nucleobases and nucleosides on cellulose layers, *J. Chromatogr.*, 268, 311, 1983.
39. **White, A. A. and Zenser, T. V.,** Separation of cyclic 3′,5′ nucleoside monophosphates from other nucleotides on aluminum oxide column, *Anal. Biochem.*, 41, 372, 1971.
40. **Holliss, D. G., Humphrey, S. M., Morrison, M. A., and Seelye, R. N.,** Reversed phase HPLC for rapid comprehensive measurement of nucleotides, nucleosides and bases of the myocardial adenine pool, *Anal. Lett.*, 17(B18), 2047, 1984.
41. **Helfferich, F. G.,** Ligand exchange: a novel separation technique, *Nature,* 189, 1001, 1961.
42. **Davankov, V. A. and Semechkin, A. V.,** Ligand exchange chromatography, *J. Chromatogr.*, 141, 313, 1977.
43. **Walton, H. F.,** Chromatography of non-ionic organic compounds on ion exchange resins, *Sep. Purif. Methods,* 4, 189, 1975.
44. **Walton, H. F.,** *Ion Exchange and Solvent Extraction,* Marinsky, J. A. and Marcus, Y., Eds., Marcel Dekker, New York, 1973, 121.
45. **Sigel, H.,** Nucleic base-metal ion interactions. Activity of N(1) or N(3) proton in binary and ternary complexes of Mn^{2+}, Ni^{2+} and Zn^{2+} with the 5′-triphosphate of inosine, guanosine, uridine and thymidine, *J. Am. Chem. Soc.*, 97, 3209, 1975.
46. **Smith, R. M. and Martell, A. E.,** *Critical Stability Constants, Vol.* 2, Plenum Press, New York, 1975.
47. **Sigel, H.,** Comparison of the stabilities of binary and ternary complexes of divalent metal ions with the 5′-triphosphates of adenosine, inosine, guanosine, cytidine, uridine and thymidine, *J. Inorg. Nucl. Chem.*, 39, 1903, 1977.

SEPARATIONS OF NUCLEIC ACID COMPONENTS BY BIDIMENSIONAL HPLC

Andreas Rizzi

INTRODUCTION

The analysis of nucleic acid components involves the separation of the purine and pyrimidine bases, the corresponding nucleosides and deoxynucleosides (only one of them in the case of uracil and thymine), the mono-, di-, triphosphates, and the cyclic phosphates. It may also include uric acid and the methyl-substituted purines and pyrimidines which are elevated in certain disease states. These compounds have to be determined either in physiological fluids or cellular extracts. A simultaneous separation and determination of all these compounds is rarely necessary in clinical and biochemical studies. However, if the number of analytes to be determined is still high or if the sample matrix is rather complex, multidimensional chromatography is specially suited to solve this separation problem.

Since its introduction by Huber in 1973,[1] multicolumn liquid chromatography is becoming increasingly recognized as a useful technique for the analysis of complex mixtures. This method offers several advantages. First, it allows the division of the overall separation problem into subproblems which can be optimized independently from each other. This is advantageous if one is interested in the simultaneous determination of many analytes. Second, a combination of columns with different retention mechanisms (e.g., separation on a reversed-phase column according to the "lipophilic" behavior, and separation on an ion-exchange column according to the strength of the electrostatic interaction) allows one to obtain an enhanced and very high separation power. Further advantages of multicolumn chromatography include trace enrichment and on-line sample cleanup by column switching. The availability of better and more versatile equipment enables the exploitation of additional features of multicolumn systems. If the apparatus is equipped with two pumps, washing and reconditioning of one column can be carried out simultaneously with the chromatographic separation on the other column,[2] thus allowing a reduced analysis time.

Contemporary multicolumn chromatography makes it possible to optimize the mobile phases for each column almost independently from the mobile phase composition of the other column(s). However, this cannot be done perfectly independently, because in general, at least a fraction of the volume of the eluents of the first column is switched to the next column in order to transfer the analytes. When designing a multicolumn system, the possible limitations resulting from this mixing of mobile phases should be kept in mind. One has to take care that any undesirable deconditioning of the subsequent column is avoided, and that the mobile-phase programming is done as independently as possible for each column.

In the forthcoming sections we will discuss the potential of dual-column chromatography in the analysis of nucleic acid components with respect to two important aspects: first, the different separation mechanisms utilized on combined columns in order to increase the separation power, and second, the adjustment of the mobile-phase composition to the requirements of subsequent columns.

SELECTION OF MULTICOLUMN SYSTEMS

Types of Stationary Phases and Separation Principles

A selection of an appropriate multicolumn system begins with a consideration of the retention mechanisms characteristic for the various types of stationary phases used for the separation of nucleobases, nucleosides, and nucleotides and of the compatibility of these phases.

Reversed-Phase Columns

Using alkyl-modified silica materials, the retention is determined by the solubility of the analytes in the mobile phase and by the adsorption equilibrium constant of the analyte on the alkyl-modified surface relative to that of the solvent components. These materials have been used successfully for the separation of bases and nucleosides and many elution programs have been published for different separation problems (cf. previous chapters of this book). Nucleotides are less frequently analyzed on these columns due to the low retention of most of these compounds (which results in insufficient resolution). However, it is possible to separate in this way the majority of nucleotides as a group from the more retained bases and nucleosides. The influence of the various parameters of the mobile phase and the packing materials on retention of nucleic acid components has been the subject of several investigations (cf. previous chapters and References 3 and 4).

Ion-Exchange Columns

Using ion-exchange columns the interaction between positively and negatively charged sites in the molecule and on the surface of the packing plays a predominant role. Cation-exchange chromatography is commonly applied to the separation of bases and nucleosides, and anion-exchange chromatography to the separation of nucleotides. Ion-exchange materials with charged groups chemically bonded to hydrophobic matrices (e.g., ion-exchange resins) may also exhibit a ''hydrophobic adsorption''. Thus, a mixed retention mechanism may be encountered with these phases. Ion-exchange materials with hydrophilic surfaces (e.g., modified silica materials) show a more unique separation mechanism.

Ion-Pair Chromatography

Ion-pair chromatography can be understood as a kind of solvent-generated ion-exchange chromatography (IEC). Therefore, the same separation principle is operative as in ion-exchange with chemically bonded groups.

Affinity Chromatography

Affinity chromatography uses a selective reaction of a special structural unit, which is immobilized on the packing material, with a selected type of compound. Within the group of nucleic acid components, affinity chromatography under low-pressure[5-10] and under high-pressure conditions[11] has been successfully used for a group separation of *cis*-diol compounds on immobilized borate in alkaline solution.[5-9,11] Ribonucleosides, 5′-ribonucleotides, and ribonucleic acids are retained on this surface in alkaline medium. Under acidic conditions, the chemical bond is broken and the compounds retained are then easily eluted. A high-performance chromatographic separation of nucleosides and 5′-nucleotides under isocratic conditions has been reported by Glad et al.[12]

Relatively few separations on unmodified silica have been reported.[13,14] This author does not know of any investigations using chemically bonded cyano-, nitro-, diol-phases. However, chemically modified stationary phases which allow complex formation with nucleotides and metal ions have been investigated.[15,16]

In using multicolumn techniques, the highest separation power is achieved by combining columns with different retention mechanisms. These combinations have the highest ''dimension'' in the sense of uncorrelated information. As outlined above, the various groups of nucleic acid components are retained on the above-mentioned stationary phases by different mechanisms. The high separation power inherent to multidimensional chromatography can therefore be utilized to a very large extent for the separation of these compounds.

Mobile-Phase Adjustment

In addition to the question of maximum resolution obtainable by the stationary phases, one also has to consider the way in which the columns should be combined on-line to avoid problems arising from the mobile phase carry-over. A principal requirement for on-line combination of columns is the complete miscibility of the mobile phases used. Silica columns, when operated with nonpolar mobile phases, cannot be combined in a simple way on-line with other types of columns employing aqueous mobile phases. Generally, for an optimal practical application one should keep in mind the following remarks:

- In order to achieve the peak-sharpening effect at the top of the second column which is needed for a good efficiency of the subsequent separation, the elution strength of the eluent designed for the second column should be greater than that of the solvent transferring the analytes from the first column.
- In order to avoid an undesirable deconditioning of the surface layers in the second column, the eluent strength of the solvent transferred from the first column must not be greater than that of the solvent used in the second column.

ON-LINE TWO-DIMENSIONAL HPLC APPLICATIONS

Although multicolumn chromatography is a potentially powerful method for analyzing purines and pyrimidines, only a small number of on-line applications discussed below has been described until now.

On-Line Combination of Affinity with Reversed-Phase Chromatography

On-line combination of a high-performance affinity chromatographic column with a reversed-phase system has been used successfully by Hagemeier et al.[11] for the determination of nucleosides and their methylated analogs in body fluids. The packing material of the first column consisted of boronic acid chemically immobilized on a silica surface (*m*-aminophenylboronic-acid silica), appropriate for the use in high-performance chromatography.[17] The selective retention of *cis*-diols in alkaline medium (pH 8.3) was used for the separation of the nucleosides from the bases, 2-deoxy-nucleosides, and all other unretained compounds. An aqueous 0.1 *M* $NH_4H_2PO_4$ solution served as a mobile phase. If the pH value is changed to 3.5, the *cis*-diols are eluted by a 0.15 *M* aqueous ammonium formate solution and transferred in a small volume (about 0.7 mℓ) to the top of the second column packed with octadecylsilica. The separation is then achieved by an elution program with increasing concentrations of methanol. Chromatograms obtained by this procedure are shown in the contribution of Hagemeier et al. in this book. A very good separation of nucleosides in serum samples is obtained. Additional UV-absorbing *cis*-diols may be found in urine; this must be taken into account when optimizing the separation on the reversed-phase column. The reported case illustrates the use of the first column for a selective group separation. The peak capacity of this first column has not been fully utilized, because there was no need for it. The high-capacity factors which can be obtained for *cis*-diols under alkaline conditions on a boronate-silica column give rise to enrichment effects when large injection volumes are used for the determination of very low sample concentrations.

On-Line Combination of Reversed Phase Chromatography with Anion-Exchange Chromatography

An example of this on-line combination has been described by Lang and Rizzi.[18] In this arrangement, the first of the two columns is filled with alkyl-silica material. Elution programs typically used with alkyl-silica columns give rise to elution of most of the nucleotides at the very beginning of a chromatogram, due to the high solubility of phosphates in water at pH values above 3, and thus allow a preseparation of these nucleotides from the stronger retained bases and nucleosides. A small fraction of about 2.5 mℓ at the beginning of the

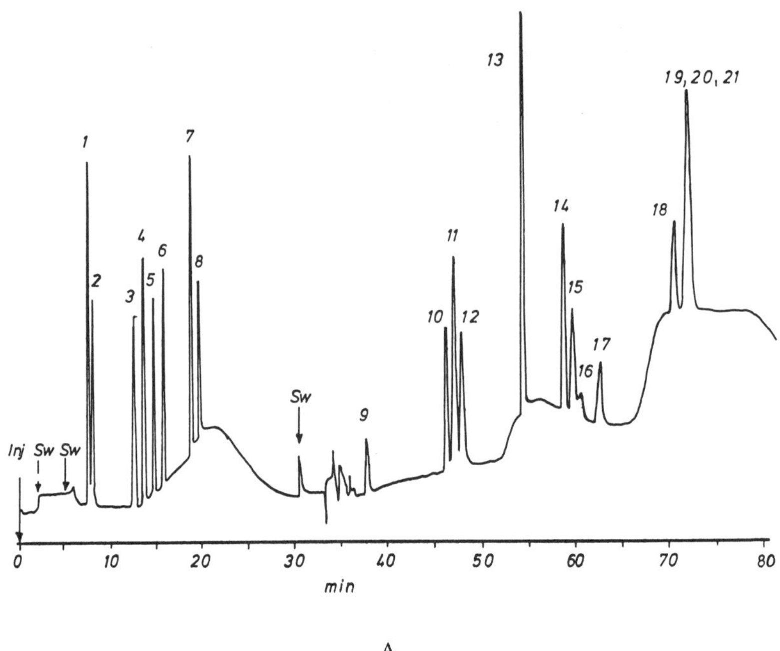

A

FIGURE 1. Bidimensional HPLC combining an octadecyl-silica column with a subsequent anion-exchange column. (A) Chromatogram of a synthetic mixture containing purine bases, nucleosides, and nucleotides. (1: Hypoxanthine, 2: xanthine, 3: inosine, 4: guanosine, 5: 2′-*d*-inosine, 6: 2′-*d*-guanosine, 7: adenosine, 8: 2′-*d*-adenosine, 9: uric acid, 10: IMP, 11: AMP, 12: GMP, 13: XMP, 14: IDP, 15: ADP, 16: GDP, 17: XDP, 18: ITP, 19: ATP, 20: GTP, 21: XTP.) (B) Chromatogram of ultracentrifuged lyzed human erythrocytes, grown in a medium spiked with adenosine (final concentration 100 μ*M*) and incubated for 5 min. Subsequent preparation for analysis (treatment with perchloric acid, extraction, and centrifugation) as described in Lang and Rizzi.[18] Same symbols used as in (A). Chromatographic conditions: column I (ODS): elution program starts with an isocratic region of about 6.5 min followed by an approximately linear gradient of increasing methanol concentration up to 20%. A small eluent fraction (zone between the first two "Sw" arrows) of about 2.5 mℓ is switched from the first to the second column; column II (AX): gradient elution program with increasing concentration of K_2HPO_4 (pH 4.75) from 0.02 to about 0.50 *M*. Elution of the AX-column starts at the third "Sw" arrow. A detailed description of the elution program and of all switching events can be found in Lang and Rizzi. (Reproduced from Lang, H. R. M. and Rizzi, A., *J. Chromatogr.*, 356, 115, 1986. With permission.)

chromatogram is transferred to the second column, which is filled with anion-exchange material. This fraction contains all tri- and diphosphates, most of the monophosphates, orotic acid, and uric acid. The nucleotides not contained in this fraction are the cyclic phosphates and sometimes AMP and TMP, depending on the reversed-phase material used. After switching the nucleotides to the anion-exchange column, the bases, nucleosides, and deoxynucleosides (and the remaining cyclic phosphates and eventually AMP and TMP) are separated on the alkyl-silica column using an appropriate elution program of increasing methanol or acetonitrile concentration. These compounds are eluted from the first column directly into the detector. After finishing the separation on the reversed-phase column, the second column is eluted using a phosphate-buffer gradient of increasing ionic strength to separate the nucleotides and the acids. If the sample contains the pyrimidine bases cytosine and uracil, as well as cytidine, these compounds may be transferred with the nucleotides to the anion-exchange column by the procedure described. Figure 1 illustrates the separations obtained by this method.

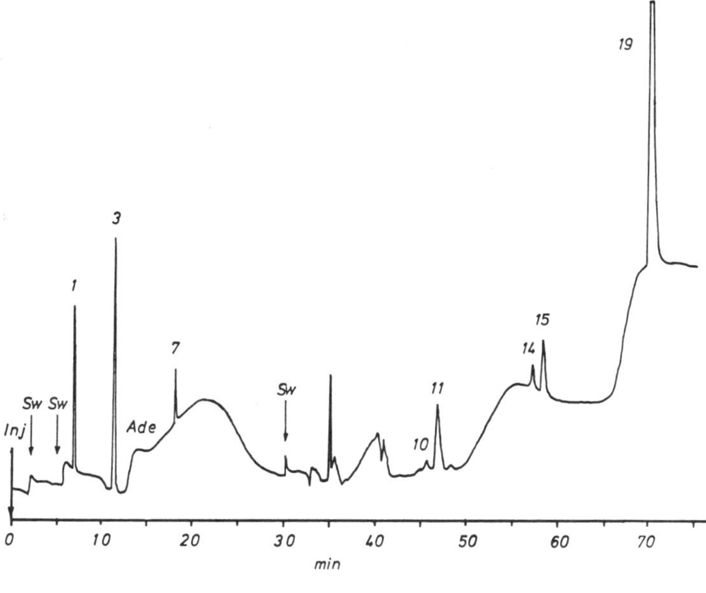

FIGURE 1B

This column-switching technique allows one to establish simultaneously the profile of the most important bases, nucleosides, and nucleotides. This can be of interest in studying cellular metabolism. An important feature of this approach is the possibility of subdividing the overall separation problem into one separation problem for bases and nucleosides and another for nucleotides. For these subproblems, many different and well-established elution programs have already been developed for single-column chromatography. They can be transposed to the two-column approach, since the mobile phases belonging to the individual columns can be optimized almost independently from each other for special problems. If one is only interested in a limited number of compounds, the elution programs can also be simplified and optimized for specific applications.

An on-line column combination in the reversed order, i.e., anion-exchange column followed by a reversed-phase column, proves to be also a powerful strategy for the separation of a wide spectrum of nucleic acid components. This presupposes that the anion-exchange material allows a preseparation and a transfer of the whole group of bases and nucleosides to the second column without a significant addition of organic solvent to the aqueous buffer solution. In this case, the mono-, di-, and triphosphates are separated in this order on the first column by increasing the ionic strength, whereas the switched bases and nucleosides can be analyzed on the second column by increasing the organic modifier concentration. When using ion-exchange materials also containing some hydrophobic adsorption sites (e.g., ion-exchange resins), the elution of bases and nucleosides may require the addition of an organic solvent. In this case, this column order may cause a deconditioning of the second column by the switched eluent fraction, since the elution program for the second column (reversed-phase material) usually starts with very small concentrations of organic solvents. Nevertheless, these materials can be used also in the mentioned column sequence if the separation problem does not require the preseparation of the whole group of bases and nucleosides.

PROSPECTS FOR FURTHER APPLICATIONS

Since only a very limited number of on-line column combinations has been investigated until now, this review will conclude with a discussion of further promising on-line column combinations for the separation of nucleic acid components which should be realized in the near future.

First, the use of ion-pair chromatography in a two-column technique is under investigation at this time.[19] Alkylamines, which are adsorbed from the mobile phase onto an alkyl-silica surface, are positively charged at pH values below 10, thus introducing anion-exchange characteristics. A disadvantage of this method is the low stability of the adsorbed amine layers with gradient programs of increasing content of organic solvents. No problems were encountered, however, in applying this method to isocratic on-line combinations, provided the mobile phases are compatible with respect to the organic solvent concentrations. This type of column can be used, for instance, after a reversed-phase column to improve the separation of insufficiently resolved analytes (e.g., hypoxanthine, xanthine, AMP, NAD, NADP with certain stationary phase materials). The elution of these compounds from the second column is done after the separation of other bases and nucleosides on the first column.

Second, several off-line combinations of columns for the separation of nucleic acid components have been described in the literature.[5-10,13,20] In the case of off-line preseparation procedures by means of ion-exchange columns (HPLC or classical LC),[13] a possible analogous on-line variant has already been discussed. The same holds for the off-line prefractionations by means of affinity chromatography.[5-10] The off-line combination of two reversed-phase columns described in the literature[21] should also be easily realized in an on-line mode. However, this type of combination cannot take advantage of the possibilities offered by multidimensional procedures with respect to selectivity programming. Nevertheless, such a combination may be useful in those cases where the first column serves mainly as a precolumn or as a gel-permeation column for the preseparation of macromolecular constituents. The usual off-line deproteinization in the analysis of cell materials can thus be avoided. The substitution of the off-line preseparation of macromolecules by an on-line size-exclusion technique seems to be very promising for many problems in nucleic acid research. This method has already been tested successfully for other compounds,[22] and one may expect that its importance will increase in the near future. This is due to the fact that the recovery of on-line methods is usually very high, since the loss by adsorption during sample pretreatment is circumvented.

REFERENCES

1. **Huber, J. F. K., Van der Linden, R., Ecker, E., and Oreans, M.,** Column switching in high-pressure liquid chromatography, *J. Chromatogr.*, 83, 267, 1973.
2. **Huber, J. F. K., Lamprecht, G., Maurer, M., Werner, E. R., Fuchs, D., Hausen, A., Reibenegger, G., and Wachter, H.,** Submitted to *J. Clin. Chem.*
3. **Scoble, H. A. and Brown, P. R.,** Reversed-phase chromatography of nucleic acid fragments, in *High-Performance Liquid Chromatography, Advances and Perspectives,* Vol. 3, Horvath, Cs., Ed., Academic Press, New York, 1983, 1.
4. **Zakaria, M., Brown, P. R., and Grushka, E.,** Mechanisms of retention in reversed phase liquid chromatographic separation of ribonucleotides and ribonucleosides and their bases, *Anal. Chem.*, 55, 457, 1983.
5. **Rosenberg, M., Wiebers, J. L., and Gilham, P. T.,** Studies on the interactions of nucleotides, polynucleotides, and nucleic acids with dihydroxyboryl-substituted celluloses, *Biochemistry,* 11/19, 3623, 1972.
6. **Uziel, M., Smith, L. H., and Taylor, S. A.,** Modified nucleosides in urine: selective removal and analysis, *Clin. Chem.*, 22, 1451, 1976.
7. **Davis, G. E., Suits, R. D., Kuo, K. C., Gehrke, C. W., Waalkes, T. P., and Borek, E.,** High-performance liquid chromatographic separation and quantitation of nucleosides in urine and some other biological fluids, *Clin. Chem.*, 23/8, 1427, 1977.
8. **Davis, G. E., Gehrke, C. W., Kuo, K. C., and Agris, P. F.,** Major and modified nucleosides in t-RNA hydrolysates by high-performance liquid chromatography, *J. Chromatogr.*, 173, 281, 1979.

9. **Gehrke, C. W., Kuo, K. C., Davis, G. E., Suits, R. D., Waalkes, T. P., and Borek, E.,** Quantitative high-performance liquid chromatography of nucleosides in biological materials, *J. Chromatogr.*, 150, 455, 1978.
10. **Schlimme, E. and Boos, K. S.,** Selective Charakterisierung von N^1-Methyladenosin neben N^7-Methylguanosine im Harn, *J. Clin. Chem. Clin. Biochem.*, 19, 55, 1981.
11. **Hagemeier, E., Kemper, K., Boos, K. S., and Schlimme, E.,** On-line high-performance liquid affinity chromatography — high-performance liquid chromatography — analysis of monomeric ribonucleoside compounds in biological fluids, *J. Chromatogr.*, 282, 663, 1983.
12. **Glad, M., Ohlson, S., Hansson, L., Mansson, M.-O., and Mosbach, K.,** High performance liquid-affinity chromatography of nucleosides nucleotides and carbohydrates with boronic acid-substituted microparticulate silica, *J. Chromatogr.*, 200, 254, 1980.
13. **Evans, J. E., Tieckelmann, H., Naylor, E. W., and Guthrie, R.,** Measurement of urinary pyrimidine bases and nucleosides by high-performance liquid chromatography, *J. Chromatogr.*, 163, 29, 1979.
14. **Hagemeier, E., Bornemann, S., Boos, K. S., and Schlimme, E.,** High-performance liquid chromatographic method for separation of dinucleotides, *J. Chromatogr.*, 237, 174, 1982.
15. **Grushka, E. and Chow, F. K.,** High-performance liquid chromatography of nucleotide-Mg(II) complexes, *J. Chromatogr.*, 199, 283, 1980.
16. **Hubert, P. and Porath, J.,** Metalchelate affinity chromatography. II. Group separation of mono- and dinucleotides, *J. Chromatogr.*, 206, 164, 1981.
17. **Hagemeier, E., Boos, K. S., Schlimme, E., Lechtenbörger, K., and Kettrup, A.,** Synthesis and application of a boronic acid substituted silica for high-performance liquid affinity chromatography, *J. Chromatogr.*, 268, 291, 1983.
18. **Lang, H. R. M. and Rizzi, A.,** Separation of purine bases, nucleosides and nucleotides by a column-switching technique combining reversed-phase and anion-exchange high-performance liquid chromatography, *J. Chromatogr.*, 356, 115, 1986.
19. **Rizzi, A.,** On-line two-dimensional separation of nucleic acid components by reversed-phase and ion-pair chromatography, in preparation.
20. **Pfadenhauer, E. H. and Sun-De Tong,** Determination of inosine and adenosine in human plasma using high-performance liquid chromatography and a boronate affinity gel, *J. Chromatogr.*, 162, 585, 1979.
21. **Taylor, G. A., Dady, P. J., and Harrap, K. R.,** Quantitative high-performance liquid chromatography of nucleosides and bases in human plasma, *J. Chromatogr.*, 183, 421, 1980.
22. **Huber, J. F. K., Lang, H. R. M., Fuchs, D., Hausen, A., Lutz, H., Reibenegger, G., and Wachter, H.,** Assay of neopterin in serum by means of HPLC applying on-line deproteinization, in *Biochemical and Clinical Aspects of Pteridines*, Vol. 3, Pfleiderer, W., Curtius, H.-Ch., and Wachter, H., Eds., Walter de Gruyter, Berlin, 1984, 81.

SIMULTANEOUS ANALYSIS OF NUCLEOTIDES, NUCLEOSIDES, AND BASES BY ION EXCHANGE AND ON-LINE ION-EXCHANGE/REVERSED-PHASE CHROMATOGRAPHY

Ardesio Floridi, Carlo A. Palmerini, and Carlo Fini

INTRODUCTION

The analysis of low molecular weight derivatives of purine and pyrimidine bases was first performed by conventional chromatography on open columns and is currently done by HPLC. Most of these procedures have been successfully applied to the separation of the bases and nucleosides or the nucleotides alone.[1,2] However, the achievement of a simultaneous profile of bases, nucleosides, and nucleotides from cellular extracts remains a somewhat troublesome analytical problem and rather few reports on this topic can be found in the literature. The analytical drawbacks arise not only from the large number of UV-absorbing compounds, which can be extracted from the cell, but also from the different physicochemical nature of the purine and pyrimidine derivatives; this causes difficulties in setting up a chromatographic system suitable for their complete and simultaneous separation.

Purine and pyrimidine bases and their nucleosides behave as amphoteric molecules. Thus, they can be separated by anion-exchange chromatography at pH values above the pK_a values of their weakly acidic groups.[3-5] Between the pK_a values of the basic and acidic groups, purine and pyrimidine bases and their nucleosides are not dissociated and they can be successfully separated by reversed-phase HPLC with isocratic[6] or gradient[7] elution modes. On the contrary, nucleotides behave as strong acids: above pH 2, the primary phosphoric group is dissociated, providing one negative charge to the nucleoside monophosphates, two negative charges to nucleoside diphosphates, and three negative charges to nucleoside triphosphates. In addition, above pH 7, all the nucleotides assume an additional negative charge due to the dissociation of the secondary phosphoric group. Anion-exchange chromatography has been successfully applied to the separation of nucleotides by conventional chromatography on polystyrene-resin-packed open columns,[8-11] or by HPLC employing either anion-exchange pellicular resins[12] or microparticulate chemically bonded strong anion-exchangers.[13]

Complex mixtures containing bases, nucleosides, and nucleoside mono- and polyphosphates are obtained when the alcohol- or the acid-soluble pools of cells and tissues are isolated. At present, an efficient simultaneous separation of these molecules can be achieved only by anion-exchange liquid chromatography. The first very important applications in this field were those reported by Anderson[10] and by Anderson et al.,[14] who used a polystyrene porous ion exchanger. These authors described the preparation of a homogeneous stationary phase, the types of mobile phases, the gradient systems, and the detection method required to obtain a satisfactory separation of complex mixtures of nucleic acid constituents. A practical application of the improvements obtained by Anderson was the separation of the acid-soluble compounds from the cellular pool, reported by Burger.[15] The separation was achieved on a 160 × 0.9-cm column packed with an AG 1 × 8 resin, 36 to 72 μm, and a linear gradient from 0.15 to 3 *M* acetate, pH 4.4. The analytical system allowed the complete separation of purine and pyrimidine bases in the presence of their respective nucleosides and nucleoside mono- and polyphosphates within 30 hr. The separation was achieved by a mechanism which only partially involved anion exchange. In fact, at pH 4.4, neither the bases nor their respective nucleosides bear any negative charge, yet the amino group at C4 of cytosine begins to be protonated. The separation of these neutral analytes takes place through a partition mechanism due to the strong nonpolar affinity shown by the purine and pyrimidine rings for the benzenoid matrix of the polystyrene ion-exchange resin.

Although allowing satisfactory resolution of complex mixtures of low molecular weight constituents of nucleic acids from tissues, the method shows the usual limitations of low-pressure open-column liquid chromatography. Thus, the time required to elute the most strongly adsorbed nucleotides is of the order of several days, and GTP cannot even be eluted. Moreover, because of the large elution volumes, the detectability afforded by this method is unsuitable for trace work.

HIGH-PERFORMANCE LIQUID ION-EXCHANGE CHROMATOGRAPHY

The difficulties encountered in open-column chromatography of complex mixtures of the bases, nucleosides, and nucleotides on anion-exchange polystyrene resins were overcome in the late 1970s by the introduction of microparticulate (particle size <20 μm) anion-exchange resins. One of the first HPLC applications of the complete separation of bases, nucleosides, and nucleotides was reported by Floridi et al.[16] The simultaneous separation of the three classes of analytes was obtained using an Aminex® A-14 resin (20 μm particle size). Compared to the open-column chromatography, this HPLC method using chloride as an anionic modifier enabled shorter analysis time, about one tenth of the time needed in Burger's method. This competing ion shows a much greater affinity than acetate for the functional groups of the resin, thus, allowing low elution volumes of analytes without serious tailing of the peaks. However, this analytical procedure requires a pH above 9 to obtain sufficient retention of the bases and nucleosides on the ion-exchange resin.

Using Aminex® A-14 resin, the optimum pH appears to be about 9.9. At this pH value, all the bases and the nucleosides, except Cyt, Cyd, and Ado, are in the anionic form and they are practically all separated on an anion exchanger because of the different pK_a values of their enolic groups. The effect of nonpolar interactions with the benzenoid matrices helps to determine the relative positions of the bases and nucleosides in the elution profile. The elution consists of a linear gradient of both ionic strength (from 0.1 to 0.4 *M* NaCl) and pH (from 9.9 to 10) and allows the complete resolution of all the three classes of compounds in approximately 3.5 hr at 55°C on a 50 × 0.6-cm column with a total gradient volume of 400 mℓ. The short analysis time and the ease of column regeneration (10 mℓ 0.3 *M* HCl followed by a 30-min washing with the first buffer) enable the use of this system in routine analyses of cell or tissue extracts.

After the complete separation of bases, nucleosides, and nucleotides was achieved by a single HPLC system with a porous polystyrene resin, other suitable techniques were designed using the Aminex® A-25 anion exchanger. This resin has a smaller particle size and a higher degree of cross-linking (8%) than Aminex® A-14 (4%). Bakay et al.[17] used Aminex® A-25 to simultaneously separate and determine the purine bases, ribonucleosides, and ribonucleotides. The complete separation was obtained using a decreasing linear gradient of both sodium tetraborate (from 0.08 to 0.01 *M*) and pH (from 9.1 to 9.0) and an increasing linear gradient of NH_4Cl (from 0.05 to 0.5 *M*) in a total volume of 90 mℓ. In about 2 hr, 16 solutes were separated using a 50 × 0.2-cm stainless-steel column thermostated at 60°C. This chromatographic system, with some modifications, was later employed by Nissinen[18] for the simultaneous analysis of purine and pyrimidine bases, nucleosides, and nucleotides. This elegant and rapid separation shown in Figure 1 was obtained by a slight modification of the gradient ionic strength and pH and by adding 2.5% ethanol to the first buffer. The presence of ethanol improves the separation of bases and nucleosides by influencing the distribution coefficients of the purines. The main advantage of this HPLC system, which utilizes a conventional porous resin, is that it allows the simultaneous analysis of all the UV-absorbing compounds present in a cell extract in about 2.5 hr. Figure 2 is an additional example of the separation obtained using a conventional polystyrene resin. The chromatogram, referred to the simultaneous analysis of some bases, nucleosides, and nucleotides,

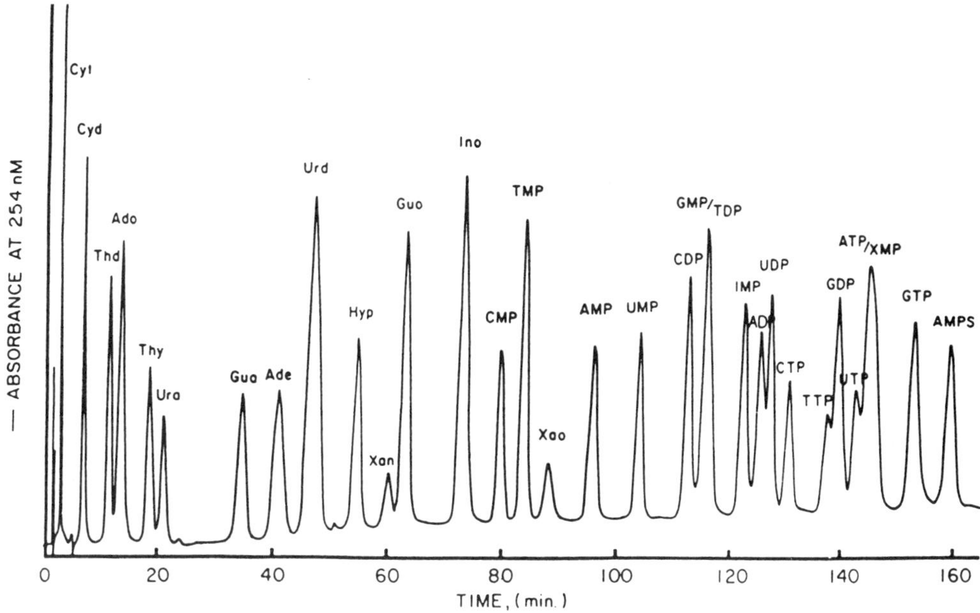

FIGURE 1. Separation of purine and pyrimidine bases, nucleosides, and nucleotides. Column: Aminex® A-25, 50 × 0.2 cm; elution: linear gradient (110 mℓ total volume) from 0.07 *M* $Na_2B_4O_y$-0.045 *M* NH_4Cl in 2.5% ethanol, pH 9.15, to 0.01 *M* $Na_2B_4O_7$-0.5 *M* NH_4Cl, pH 8.80, with 15 min gradient delay; flow rate: 0.5 mℓ/min; temperature: 65°C. (From Nissinen, E., *Anal. Biochem.*, 106, 497, 1980. With permission.)

was obtained in our laboratory using a 15 × 0.46-cm analytical column and a 0.46 × 3-cm guard column, both packed with CDR 10 resin (20% cross-linking), a quaternary ammonium anion exchanger with a polystyrene matrix of 7 μm average particle size. A satisfactory separation of all compounds, with the exception of CDP (which overlaps with GMP) and of UTP (which overlaps with ATP), was obtained in about 2 hr. The chromatography was carried out at 55°C using a linear gradient. The low strength eluent consisted of 0.035 *M* potassium tetraborate-0.025 *M* NH_4Cl, pH 9.8, and the high strength eluent was 0.01 *M* potassium tetraborate-0.25 *M* NH_4Cl, pH 9.0. The analysis time was shorter than with the polystyrene resins. Further improvements are possible by additional optimization of experimental conditions. However, all HPLC methods discussed thus far do not attain the efficiency of the silica-bonded HPLC phases. Moreover, the analyses are time consuming and the resolution is often unsatisfactory.

ON-LINE COMBINATION OF REVERSED-PHASE AND ANION-EXCHANGE HPLC COLUMNS

Over the last several years, HPLC techniques using silica-bonded phases[19-21] proved to be of great help in metabolic and clinicobiochemical studies of the low molecular weight constituents of nucleic acids. In this field, HPLC represents one of the best analytical techniques for the determination of extra- and intracellular low molecular weight metabolites, since it offers speed, sensitivity, and high efficiency. However, since the simultaneous separation of bases, nucleosides, and nucleotides can only be achieved using anion-exchange chromatography at alkaline pH values, HPLC on silica-bonded phases appears to be almost unfeasible. This limitation is due to the solubility of the silica matrices at pH values above 9.5, required for the simultaneous anion-exchange chromatography of these compounds. Moreover, at neutral or acidic pH values, neither reversed-phase nor ion-exchange or ion-pair chromatography affords the simultaneous separation of the three classes of compounds.

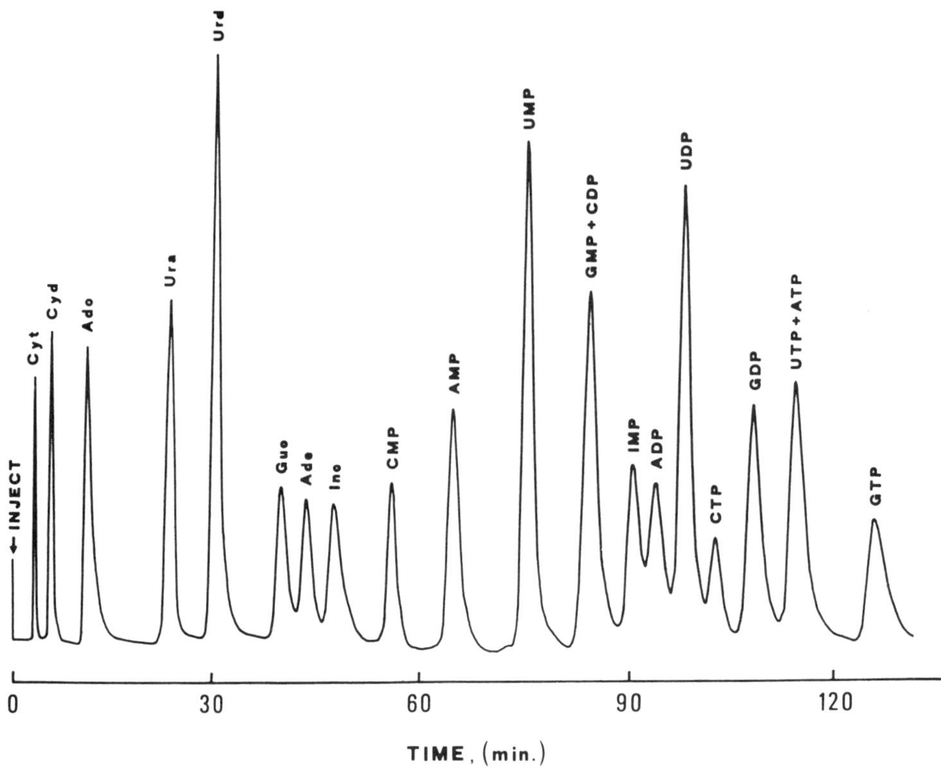

FIGURE 2. Simultaneous separation of standard bases, nucleosides, and nucleotides (1 to 2 nmol) by anion-exchange HPLC. Instrumentation: Waters chromatograph equipped with two Model 6000A solvent delivery systems, a Model 660 solvent programmer, a Model U6K sample injector, a Model 440 UV-detector, a Model 730 data module, and a thermostatic oven. Column: 15 × 0.46 cm column and 3 × 0.46 cm precolumn, both packed with CDR 10 resin; elution: linear gradient (curve 6, 180 min) from 0.035 *M* $K_2B_4O_7$-0.025 *M* NH_4Cl, pH 9.8, to 0.01 *M* $K_2B_4O_7$-0.25 *M* NH_4Cl, pH 9.0; temperature: 55°C; flow rate: 1.0 mℓ/min; detection: 254 nm at 0.1 aufs.

However, one can envisage an HPLC system utilizing simultaneously the reversed-phase chromatography for the separation of bases and nucleosides, and ion-exchange chromatography for the separation of nucleotides. This possibility has been explored in our laboratory and a new technique for the simultaneous separation of the UV-absorbing low molecular weight compounds from cell extracts has been designed. This technique utilizes on-line combination of a reversed-phase column (octadecyl) and an anion-exchange column packed with a microparticulate silica bonded to a strong anion exchanger. The separation is carried out by gradient elution and the analysis is performed at 15°C. The separation of a multi-component mixture of bases, nucleosides, and nucleotides, using the described system, is shown in Figure 3. The resolution is satisfactory and the peak shapes are symmetrical. The analytes examined are those normally present at the highest concentration in cell extracts. However, if xanthine were present in the mixture, it would overlap with hypoxanthine, and a rather broad adenine peak would elute between GMP and inosine. The separation is carried out at 15°C in order to obtain a better resolution of the first eluted nucleotide, CMP, and those bases and nucleosides which are poorly retained by the reversed-phase column.

The optimization of this chromatographic system, which couples reversed-phase to anion-exchange, consisted of adjusting the concentrations of the ionic (phosphate) and organic (methanol) modifiers, in order to avoid overlapping of the compounds adsorbed on the first column with those emerging from the second column. The separation of nucleoside mono- and polyphosphates on the anionic exchanger was achieved by using an elution system

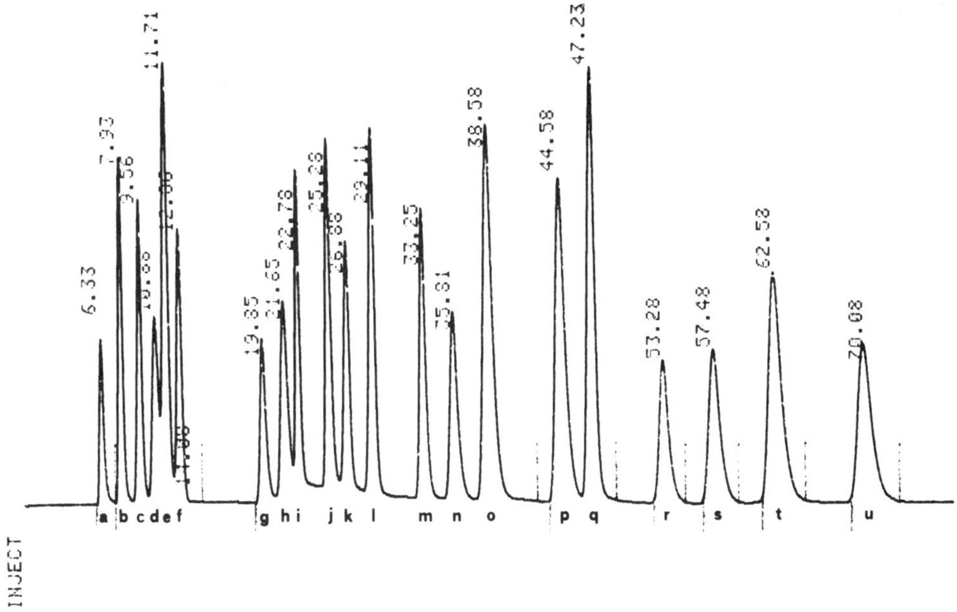

FIGURE 3. Simultaneous separation of reference bases, nucleosides, and nucleosides mono- and polyphosphates (0.5 to 1 nmol). Peak identification: a = Cyt; b = Ura; c = Cyd; d = CMP; e = Hyp; f = Urd; g = AMP; h = UMP; i = IMP; j = GMP; k = Ino; l = Guo; m = UDP; n = CDP; o = ADP; p = GDP; q = Ado; r = UTP; s = CTP; t = ATP; u = GTP. Instrumentation: as in Figure 2; columns: Spherisorb® ODS2, 15 × 0.46 cm, 5 μm, and Partisil® SAX, 25 × 0.4 cm, 10 μm; elution: linear gradient (curve 6, 110 min) from 0.01 *M* $NH_4H_2PO_4$, pH 4.4, containing 0.01% acetone, to methanol — 0.6 *M* $(NH_4)_2HPO_4$, adjusted to pH 4.4 with phosphoric acid, (20:80 v/v); temperature: 15°C; detection: 254 nm at 0.1 aufs.

employing a phosphate buffer. This system proved to be highly efficient. The elution order for the phosphorylated compounds was similar to that reported by Hartwick and Brown,[13] who used a similar ion-exchange system, but omitted the organic modifier in the high-strength buffer. The methanol percentage used in the present method ensures a good resolution of the nucleosides, which are strongly adsorbed on the reversed-phase column. The nucleosides are Ino and Guo, which are eluted after the nucleoside monophosphates, and Ado, which is eluted after the nucleoside diphosphates.

The two-column HPLC system described proves to be highly efficient and allows a satisfactory separation of the bases, nucleosides, and nucleotides in less than 70 min. The analysis time is considerably shorter than that required by the HPLC systems using conventional ion-exchange resins.

The applicability of the HPLC system developed in this laboratory was tested in the analysis of the UV-absorbing compounds in rat brain perchloric acid extracts. A typical chromatographic profile of the extract equivalent to 2 mg of fresh brain is reported in Figure 4. As can be seen, with the exception of the hypoxanthine-xanthine pair which is not resolved, each component of interest appears in the chromatogram as a sharp peak sufficiently well separated for accurate determination. Moreover, a comparison of both retention times and peak band widths of analytes from the brain extract and reference compounds shows absence of interferences caused by coeluting unknowns.

CONCLUSIONS

A great number of metabolic or clinicobiochemical studies requires simultaneous mapping of the bases, nucleosides, and nucleotides in biological samples. Currently, the most popular

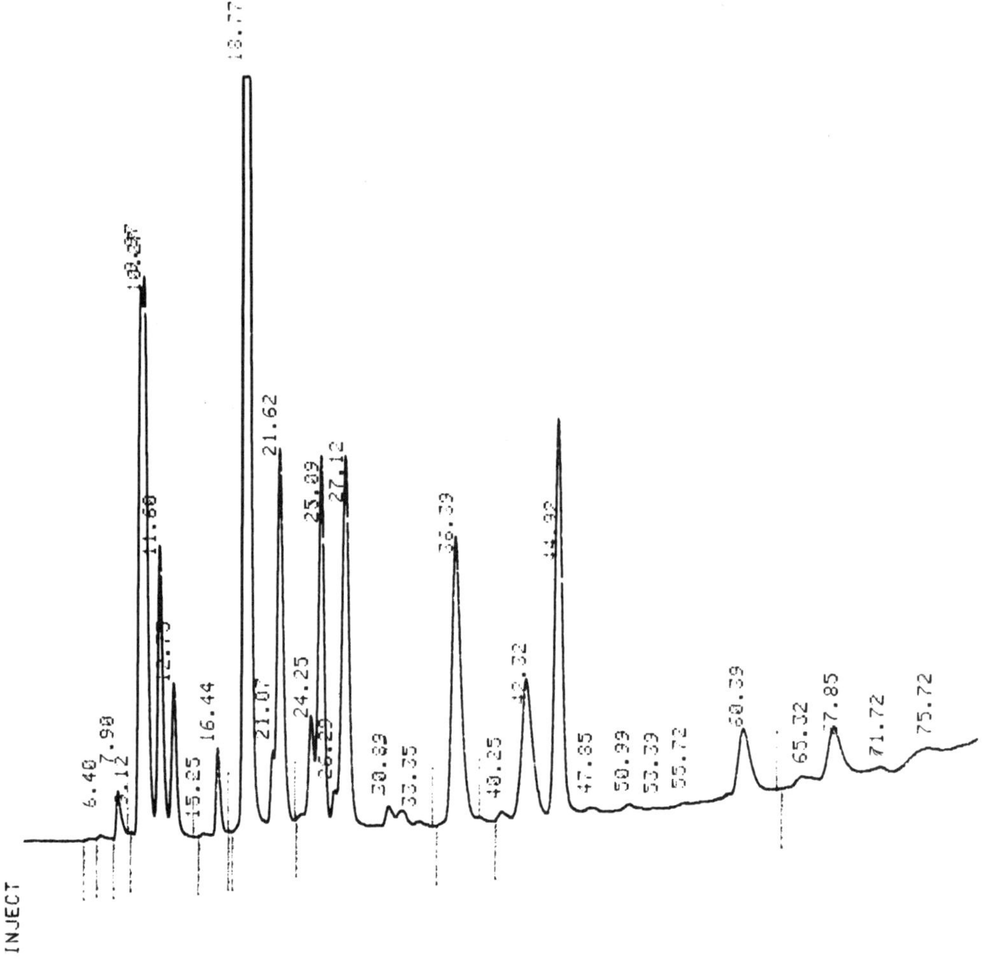

FIGURE 4. Separation of the UV-absorbing compounds from a sample (20 μℓ) of a neutralized perchloric acid extract of rat brain by coupled reversed-phase and anion-exchange HPLC. Experimental conditions: as in Figure 3.

and successful methods are based on HPLC systems employing polystyrene conventional resins and elution at alkaline pH. The separation of the more lipophilic molecules takes place by anion exchange and partition. Because of the physicochemical properties of the solutes under investigation and the chromatographic behavior of the porous polystyrene exchangers, the separation of multicomponent mixtures cannot be obtained in less than 2 hr. However, routine assays of biochemical/biomedical samples require shorter analysis times, which can be obtained using silica-bonded phases. Simultaneous assays of the bases, nucleosides, and nucleotides require the combined use of reversed-phase and anion-exchange modes, and a binary linear gradient of both an ionic modifier (to elute nucleotides) and an organic modifier (to elute bases and nucleosides). An example of such an analytical procedure is described in this report, and it is hoped that further optimization of this technique will provide improved resolution and faster analyses.

REFERENCES

1. **Brown, P. R., Krstulovic, A. M., and Hartwick, R. A.,** Current state of the art in the HPLC analyses of free nucleotides, nucleosides, and bases in biological fluids, *Adv. Chromatogr.*, 18, 101, 1980.
2. **Zakaria, M. and Brown, P. R.,** High-performance liquid column chromatography of nucleotides, nucleosides and bases, *J. Chromatogr.*, 226, 267, 1981.
3. **Schmitz, H., Hurlbert, R. B., and Potter, V. R.,** Nucleotide metabolism. III. Mono-, di-, and triphosphates of cytidine, guanosine, and uridine, *J. Biol. Chem.*, 209, 41, 1954.
4. **Singhal, R. P. and Cohn, W. E.,** Analytical separation of nucleosides by anion-exchange chromatography: influence of pH, solvents, temperature, concentration, and flow rate, *Anal. Biochem.*, 45, 585, 1972.
5. **Brown, P. R., Bobick, S., and Hanley, F. L.,** The analysis of purine and pyrimidine bases and their nucleosides by high-pressure liquid chromatography, *J. Chromatogr.*, 99, 587, 1974.
6. **Simmonds, R. J. and Harkness, R. A.,** High-performance liquid chromatographic methods for base and nucleoside analysis in extracellular fluids and in cells, *J. Chromatogr.*, 226, 369, 1981.
7. **Hartwick, R. A., Assenza, S. P., and Brown, P. R.,** Identification and quantitation of nucleosides, bases and other UV-absorbing compounds in serum, using reversed-phase high-performance liquid chromatography. I. Chromatographic methodology, *J. Chromatogr.*, 186, 647, 1979.
8. **Pontis, H. G. and Blumsom, N. L.,** A method for the separation of nucleotides by concave gradient elution, *Biochim. Biophys. Acta*, 27, 618, 1958.
9. **Ingle, J.,** A method for the improved separation of soluble nucleotides by ion-exchange chromatography, *Biochim. Biophys. Acta*, 61, 147, 1962.
10. **Anderson, N. G.,** Analytical techniques for cell fractions. II. A spectrophotometric column monitoring system, *Anal. Biochem.*, 4, 269, 1962.
11. **Virkola, P.,** An automated system for ion-exchange chromatography of acid-soluble nucleotides at the nanomole level, *J. Chromatogr.*, 51, 195, 1970.
12. **Horvath, C. G., Preiss, B. A., and Lipsky, S. R.,** Fast liquid chromatography: an investigation of operating parameters and the separation of nucleotides on pellicular ion exchangers, *Anal. Chem.*, 39, 1422, 1967.
13. **Hartwick, R. A. and Brown, P. R.,** The performance of microparticle chemically-bonded anion-exchange resins in the analysis of nucleotides, *J. Chromatogr.*, 112, 651, 1975.
14. **Anderson, N. G., Green, J. B., Barber, M. L., and Ladd, F. C.,** Analytical techniques for cell fractions. III. Nucleotides and related compounds, *Anal. Biochem.*, 6, 153, 1963.
15. **Burger, C. L.,** Measurement of nucleotide, amino acid and DNA contents of tissues after whole-body X-irradiation, *Anal. Biochem.*, 20, 373, 1967.
16. **Floridi, A., Palmerini, C. A., and Fini, C.,** Simultaneous analysis of bases, nucleosides and nucleosides mono- and polyphosphates by high-performance liquid chromatography, *J. Chromatogr.*, 138, 203, 1977.
17. **Bakay, B., Nissinen, E., and Sweetman, L.,** Analysis of radioactive and nonradioactive purine bases, nucleosides, and nucleotides by high-speed chromatography on a single column, *Anal. Biochem.*, 86, 65, 1978.
18. **Nissinen, E.,** Analysis of purine and pyrimidine bases, ribonucleosides, and ribonucleotides by high-pressure liquid chromatography, *Anal. Biochem.*, 106, 497, 1980.
19. **Brown, P. R. and Krstulovic, A. M.,** Practical aspects of reversed-phase liquid chromatography applied to biochemical and biomedical research, *Anal. Biochem.*, 99, 1, 1979.
20. **Maybaum, J., Klein, F. K., and Sadee, W.,** Determination of pyrimidine ribotide and deoxyribotide pools in cultured cells and mouse liver by high-performance liquid chromatography, *J. Chromatogr.*, 188, 149, 1980.
21. **Payne, S. M. and Ames, B. N.,** A procedure for rapid extraction and high-pressure liquid chromatographic separation of the nucleotides and other small molecules from bacterial cells, *Anal. Biochem.*, 123, 151, 1982.

ANALYSIS OF URINARY AND PLASMA NUCLEOSIDES BY COLUMN SWITCHING TECHNIQUE

Eberhard Hagemeier, Karl-Siegfried Boos, Klaus Kemper, and Eckhard Schlimme

INTRODUCTION

During the last decade evidence has been increasing that the investigation of major and minor nucleosides in human urine might be helpful in cancer diagnosis as well as in exploring and controlling the effectiveness of therapy (therapeutic drug monitoring).[1-3] Furthermore, the determination of major nucleosides in human plasma has attracted attention due to their important role both in health condition and some diseases. For example, adenosine plays a role in vasodilation, neurotransmission, immune response, and more.[4,5] The exploitation of nucleosides as marker molecules in pathobiochemistry or generally in routine clinical analysis requires the availability of sensitive, rapid, and simple analytical methods for the determination of these target compounds in biological samples.

Since more than 50 modified nucleosides have been isolated and characterized in tRNA[6,7] and more than 25 in urine,[8] investigators in this field had to cope with two serious tasks: (1) to prepurify the compounds of interest from such complex matrices as urine or serum, and (2) to analyze a large variety of structurally and chemically related compounds (Figure 1). The first was accomplished by applying low-pressure boronic acid-affinity chromatography for the group-selective removal of ribonucleosides from their matrix,[9,10] and the second by exploiting the reversed-phase (RP) mode of high-performance liquid chromatography (HPLC) for the analysis of the majority of the naturally occurring major and minor nucleosides.[10-12] This sequential, i.e., off-line, separation technique has been widely used.[2,10,13-19]

As this off-line technique includes numerous working steps, it was the authors' objective to develop an instrumentally connected, on-line multidimensional liquid chromatographic system for the direct cleanup and analysis of the target compounds.

PREFRACTIONATION AND ANALYSIS OF NUCLEOSIDES IN BIOLOGICAL SAMPLES

Cleanup of Ribonucleosides by Boronic Acid-High-Performance Liquid Affinity Chromatography (HPLAC)

In order to set up a multidimensional HPLC system,[20,21] the functional affinity ligand *m*-aminobenzeneboronic acid was coupled to silica (Figure 2).[22,23]

An analogous material for HPLAC was also used for the analytical resolution of some ribo- and deoxyribonucleosides, ribonucleotides, and saccharides by applying isocratic conditions and/or a pH gradient.[24-26] For the desired group-selective prefractionation and cleanup of urinary ribonucleosides under HPLAC conditions a simple pH step-elution was carried out.[23,27,28]

Figure 3 illustrates the mechanism and conditions for the boronic acid-HPLAC by pH step-elution.

After sample application, *cis*-diol compounds such as ribonucleosides are selectively retarded under slightly alkaline conditions, whereas the sample matrix is discharged. Ribonucleosides, selectively bound to the boronic acid ligand, are then eluted by lowering the pH of the mobile phase below 4.

Figure 4 shows the elution profiles obtained in the cleanup of urinary ribonucleosides by boronic acid HPLAC.

The high selectivity and capacity of the affinity material, as well as the reduction of

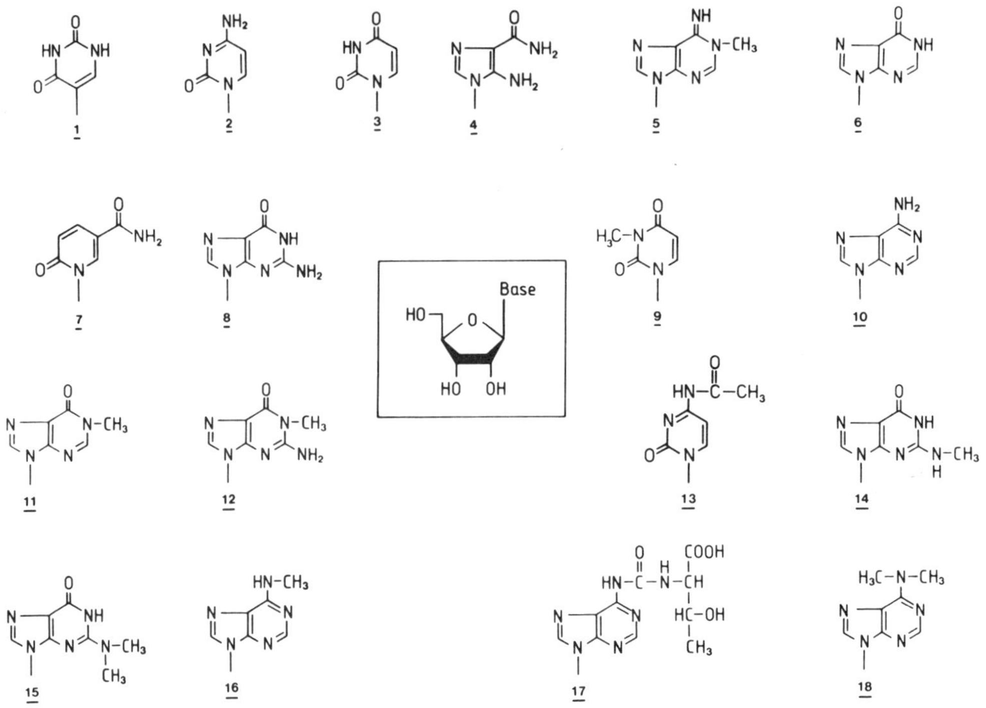

FIGURE 1. Formula of investigated ribonucleosides. (1) Pseudouridine (ψ), (2) cytidine (Cyd), (3) uridine (Urd), (4) 5-aminoimidazole-4-carboxamide-*N*-ribofuranoside (AICAR), (5) N1-methyladenosine (m^1Ado), (6) inosine (Ino), (7) 2-pyridone-5-carboxamide-*N*-ribofuranoside (PCNR), (8) guanosine (Guo), (9) N3-methyluridine (m^3Urd), (10) adenosine (Ado), (11) N1-methylinosine (m^1Ino), (12) N1-methylguanosine (m^1Guo), (13) N4-acetylcytidine (ac^4Cyd), (14) N2-methylguanosine (m^2Guo), (15) N2-dimethylguanosine (m^2_2Guo), (16) N6-methyladenosine (m^6Ado), (17) N6-(carbamoyl-threonyl)adenosine (t^6Ado), (18) N6-dimethyladenosine (m^6_2Ado).

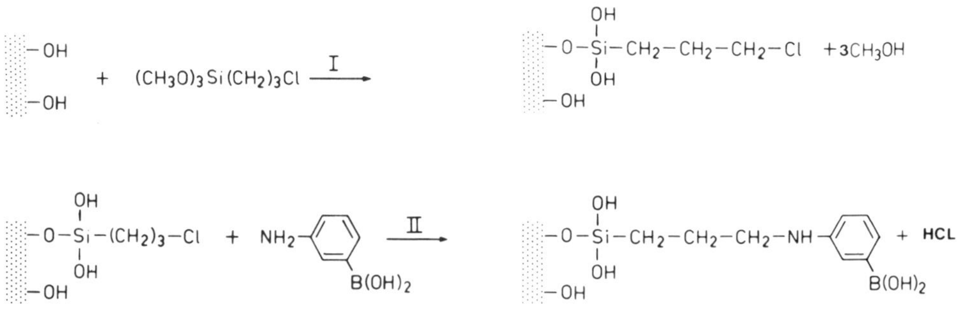

FIGURE 2. Boronic acid functionalized silica. Silica LiChrosorb® Si 100 (5 μm, Merck) was substituted under argon with γ-chloropropyltrimethoxysilane (I) and subsequently reacted with *m*-aminobenzeneboronic acid (II) according to Hagemeier et al.[22] (Taken from Hagemeier, E., Boos, K.-S., Schlimme, E., Lechtenbörger, K., and Kettrup, A., *J. Chromatogr.*, 268, 291, 1983. With permission.)

cleanup time and elution volume, formed the basis for the successful application of a column-switching technique, i.e., for the direct transfer of the acidic HPLAC-fraction onto an analytical RPLC column.

Reversed-Phase High-Performance Liquid Chromatography (RPLC) of Ribonucleosides

The RP mode of HPLC is a powerful tool for nucleoside analysis.[29-31] Figure 5 shows the RPLC separation of biologically important major and modified ribonucleosides.

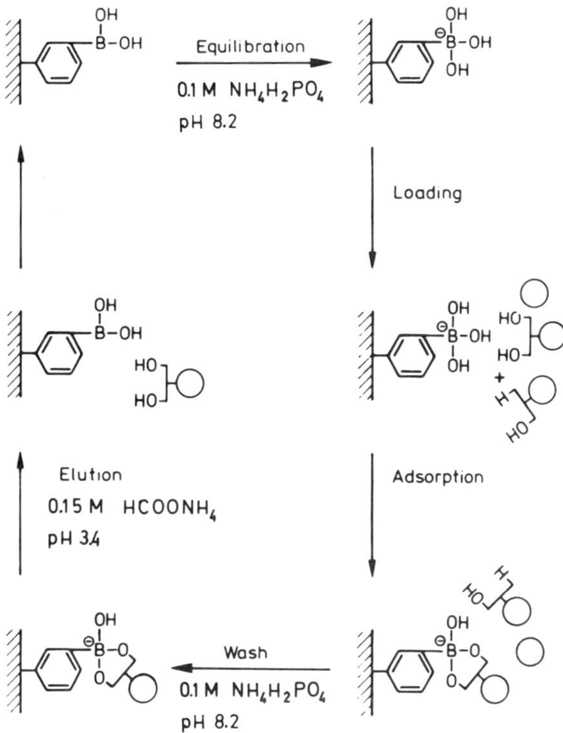

FIGURE 3. Boronic acid affinity chromatography based on pH shift. (Taken from Hagemeier, E., Kemper, K., Boos, K. S. and Schlimme, E., *J. Clin. Chem. Clin. Biochem.*, 22, 175, 1984. With permission.)

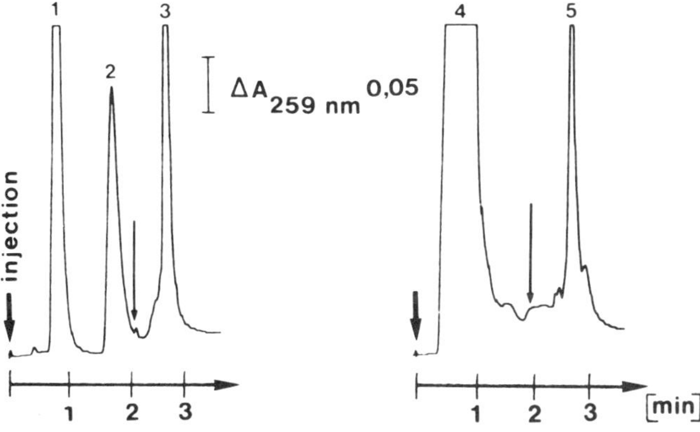

FIGURE 4. Prefractionation of urinary ribonucleosides by HPLAC Left: fractional separation of a synthetic mixture of deoxythymidine (1), deoxyadenosine (2), pseudouridine (3), and adenosine with boronic acid silica under pH-step gradient conditions. Elution sequence: 2 min 0.1 *M* ammonium dihydrogen-phosphate pH 8.2; 0.15 *M* ammonium formate pH 3.4 (indicated by an arrow); flow rate 1 mℓ/min. Detection: UV, 259 nm. Right: cleanup and group-specific separation of urinary *cis*-diol compounds from 50 μℓ membrane-filtered urine; 4 = urinary UV-absorbing compounds; 5 = urinary *cis*-diol compounds. Elution sequence as before. (Taken from Hagemeier, E., Boos, K.-S., Schlimme, E., Lechtenbörger, K., and Kettrup, A., *J. Chromatogr.*, 268, 291, 1983. With permission.)

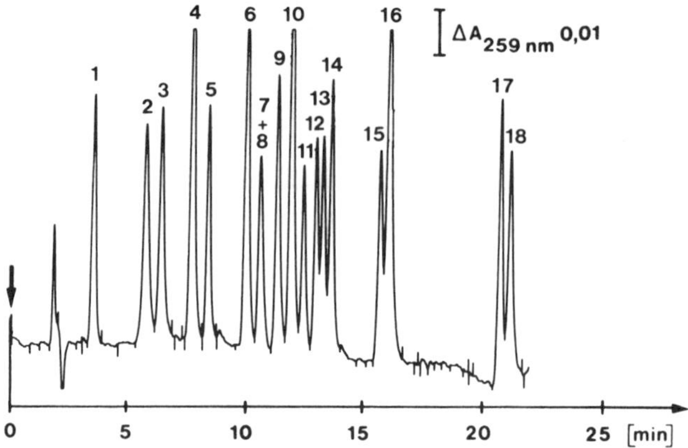

FIGURE 5. RPLC of a synthetic mixture of 18 ribonucleosides. Chromatographic conditions: LiChrosorb® RP-18 (7 μm, 250 × 4 mm ID). Detection: UV, 259 nm. Elution: 0.15 *M* ammonium formate pH 3.4; after 2 min a linear gradient up to 11% methanol in 8 min, followed by a linear gradient up to 16% methanol in 4 min. After 2 min at 16% methanol a linear gradient up to 50% methanol in 9 min; flow rate, 1.8 mℓ/min. The peak numbering refers to Figure 1. Application of 2 to 10 nmol of each nucleoside.[53]

ON-LINE MULTIDIMENSIONAL LIQUID CHROMATOGRAPHY OF NUCLEOSIDES

Instrumentation

The schematic diagram of the multidimensional chromatographic system employed is given in Figure 6. The chromatograph consists of a commercial, microprocessor-controlled two-pump gradient system, which was further equipped with a third HPLC pump, a high-pressure six-port, two-way valve for column switching, and a six-port sampling valve for sample introduction into the affinity column (column 1). This instrumentation allows the independent use of the basic gradient system besides the multidimensional mode. Thus, reliability of the overall system performance can easily be controlled by comparing the off-line RPLC analysis of a standard mixture of ribonucleosides with the on-line (HPLAC/RPLC) analysis of an adequate sample. By use of a high-pressure pneumatically operated column-switching valve and an autosampler, both controlled by time-programmable external events from the chromatograph, the entire on-line analysis of ribonucleosides can be automated. Such fully automated devices for the application in multidimensional HPLC have been described recently.[21,32,33,43]

Column 1 (30 × 4.0 mm ID) in this two-dimensional chromatographic system functions as a precolumn for the cleanup of ribonucleosides. Boronic acid-substituted silica, either purchased (dihydroxyborylpolyol Si 100, 5 μm SERVA Feinbiochemica, Heidelberg, West Germany) or laboratory prepared according to Hagemeier et al.,[22,23] is used as a stationary phase for this column. Column 2 (250 × 4.0 mm ID) filled with LiChrosorb® RP-18, 7 μm (Merck, Darmstadt, West Germany) is used as the analytical column.[27,28] Other column dimensions or RP materials, suitable for ribonucleoside analysis, may be used in analogy.

Analysis of Ribonucleosides in Urine and Serum

Figure 7 depicts the switching-valve configurations and solvent flow paths. In valve position "Load", column 1 and 2 are equilibrated with the respective alkaline or acidic buffer. After the introduction of membrane-filtered urine or ultrafiltrated serum into the HPLAC column (column 1), the affinity column is washed with alkaline buffer until the

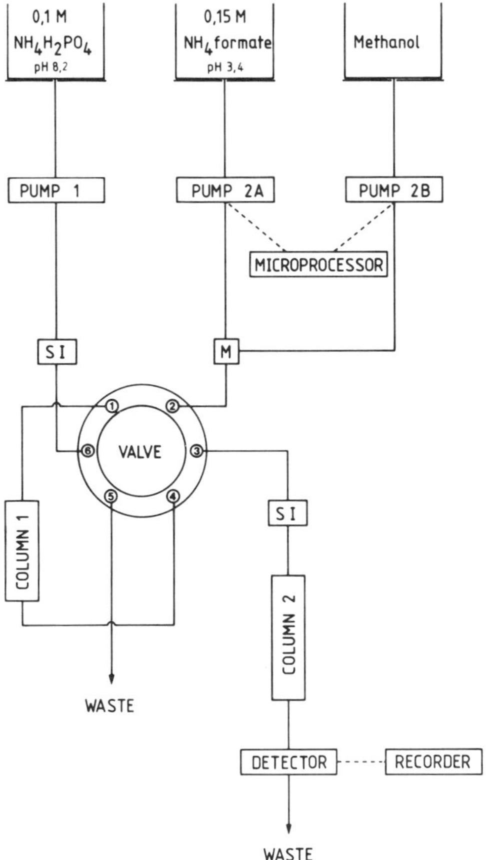

FIGURE 6. On-line system setup. (Taken from Hagemeier, E., Kemper, K., Boos, K.-S., and Schlimme, E., *J. Chromatogr.*, 282, 663, 1983. With permission.)

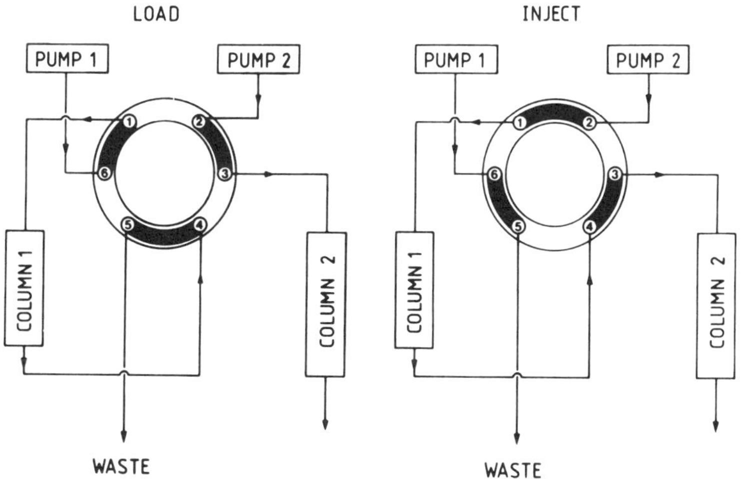

FIGURE 7. Valve-switching positions. (Taken from Hagemeier, E., Kemper, K., Boos, K.-S., and Schlimme, E., *J. Chromatogr.*, 282, 663, 1983. With permission.)

sample matrix is discharged. During the wash period, *cis*-diol compounds are selectively retarded. As retardation of the target compounds is limited by the boronic-acid capacity of the support, buffer pH, as well as column dimensions, the maximum range of time which is acceptable for complete recovery of the compounds of interest has to be optimized for the given instrumental conditions.

After this cleanup step, the valve is switched to position "Inject", whereby column 1 is series connected in front of the analytical RP column (column 2). The ribonucleosides sorbed on column 1 are then eluted under acidic conditions in a small volume through positions 2-1-4-3 of the valve and concentrated on top of column 2. The chromatographic zone transferred in this way to the analytical column corresponds to peaks 3 or 5 in Figure 3, respectively. The acidic buffer used for the elution of ribonucleosides from column 1 is also used as a mobile phase for column 2. Separation of nucleosides on column 2 is then carried out by increasing the amount of methanol in the mobile phase of column 2. After the transfer of ribonucleosides from column 1 to column 2, the switching valve is returned to position "Load", and column 1 can be reconditioned for the next analysis. The time range for the transfer step also has to be adapted to the instrumental conditions. In order to keep the elution volume of ribonucleosides from the HPLAC column at a minimum, the acidic elution buffer should fulfill the following requirements: (1) the pH should be below pH 4 to cease ligand-analyte interactions, (2) ionic strength and buffer capacity should lead to a fast attainment of an acidic state in the HPLAC column, and (3) it should be free of organic solvents to ensure enrichment of the compounds of interest on the RP analytical column.

Figure 8 shows the on-line HPLAC/RPLC separation of (1) a synthetic mixture of ribonucleosides (2) membrane-filtered native human urine, and (3) deproteinized human serum.

The comparison of Figures 5 and 8A clearly demonstrates that the column-switching technique under the conditions applied[27,28] does not essentially affect bandbroadening and resolution of the nucleosides investigated. Whereas modified nucleosides are present in human urine (Figure 8B) in much larger amounts than the major ribonucleosides, the opposite is true for human serum. Figure 8C shows the separation of the major ribonucleosides inosine, guanosine, and adenosine in human serum. The quantitative determination of major nucleosides in serum samples, however, is considerably complicated by the activity of endogenous enzymes present in native serum.[19,35,36] Thus, the sample processing procedures prior to analysis have to be improved and standardized as described by Ontyd and Schrader.[36]

Accuracy and Precision

Figure 9 shows the excellent linearity between sample volume analyzed and amount of nucleoside found. Linearity was proved for the determination of m^1Ado, m^1Ino, m^2Guo, m^2_2Guo, PCNR, and t^6Ado in 20 to 100 $\mu\ell$ of native urine. Between 100 and 4000 pmol of the respective nucleoside can be measured with correlation coefficients for linear regression between 0.998 and 0.999. The wide linear range is sufficient for routine determination of nucleosides in human urine. The reliability of the on-line HPLAC/RPLC system is illustrated in Tables 1 and 2, which summarize the results obtained for matrix-independent and matrix-dependent recoveries.

Off-Line/On-Line Comparison

Figure 10 shows the good correlation of results obtained by off-line and on-line method for the determination of m^1Ado in 25 different urines (values for m^1Ino, m^2Guo, and m^2_2Guo are given in legend to Figure 10). The on-line technique, however, is distinguished from the off-line method by the following features (compare Figure 11):

1. Total analysis time is shortened from at least 130 to 33 min.
2. Laborious and error-prone evaporation and redissolution steps are avoided, thus leading to an improvement of analytical precision.

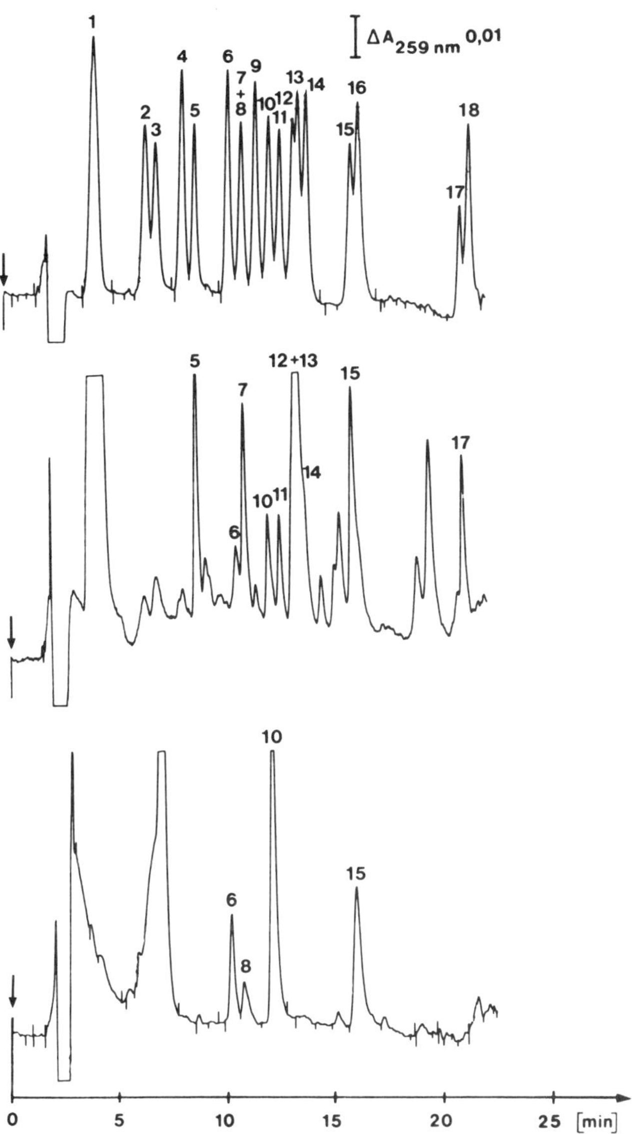

FIGURE 8. HPLAC/RPLC of ribonucleosides. (A) On-line chromatography of a 100-μℓ sample of a synthetic mixture of ribonucleosides (compare with Figure 5); (B) on-line chromatography of 10 μℓ of a membrane-filtered human urine sample; (C) on-line chromatography of 200 μℓ of a human serum ultrafiltrate. Chromatographic conditions: HPLAC-column 1 was equilibrated for 2 min in valve position "Load" with 0.1 mol/ℓ ammonium phosphate pH 8.2. After sample injection column 1 was washed with the same buffer for 2 min. After this cleanup step the valve was switched to "Inject" and, thereby, series connected in front of column 2. The group specifically bound ribonucleosides on column 1 were then transferred with 0.15 mol/ℓ ammonium formate pH 3.4 on top of column 2 over a period of 1.5 min. The valve was then switched back into position "Load" and elution of column 2 was carried out as shown in Figure 5. The peak numbering refers to Figure 1.

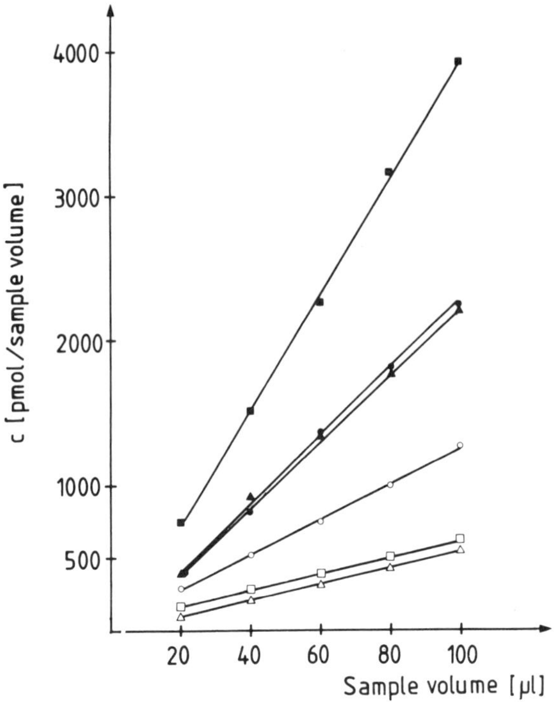

FIGURE 9. Linearity of HPLAC/RPLC analysis. m_2^2Guo (■—■), r = 0.999; m^1Ado (●—●), r = 0.999; PCNR (▲—▲), r = 0.999; m^1Ino (○—○), r = 0.998; m^2Guo (□—□), r = 0.999; t^6Ado (△—△), r = 0.999. (Taken in part from Hagemeier, E. Kemper, K., Boos, K.-S., and Schlimme, E., *J. Clin. Chem. Clin. Biochem.*, 22, 175, 1984. With permission.)

3. The analytical procedure can be easily controlled for its reliability.
4. Sensitive samples can be processed very rapidly under mild conditions.
5. Small sample volumes, as for example, micropunctuates or RNA hydrolysates, can be directly applied and analyzed.
6. The apparatus allows easy automation.

CONCLUSIONS

A chromatographic system is described which allows rapid and convenient analysis of endogenous as well as exogenous ribonucleosides in biological samples. The acquisition of a boronic acid functionalized support suitable for use in HPLC was the basic requirement for the development of a combined HPLAC/RPLC system. Table 3 summarizes the excretion values for various urinary nucleoside markers of clinical interest which were determined by HPLC and HPLAC/HPLC, respectively. As boronic acid-affinity chromatography has turned out to be useful for the analysis of various compounds of biological interest (Table 4), the high-performance mode, used solely or in combination with a column-switching technique, should also be applicable to other fields of biochemical research.

Table 1
MATRIX-INDEPENDENT RECOVERY OF NUCLEOSIDES[23,34]

Nucleoside	Recovery (%)	SD (%)
m^1Ado	99.4[a]	1.3[b]
m^1Ino	99.7	1.6
m^2Guo	97.0	3.6
m^2_2Guo	99.8	4.3
PCNR	100.2	1.6
t^6Ado	99.4	0.6

[a] Each value is an average of three assays.
[b] Relative standard deviation.

Table 2
MATRIX-DEPENDENT RECOVERY OF NUCLEOSIDES ADDED TO POOLED CONTROL URINE[23,34]

Nucleoside	Concentration[a] (μmol/ℓ) Urine	Spike	Urine spike (found)	Average recovery (%)
m^1Ado	11.49	5.17	16.41 ± 0.3[b]	99
m^1Ino	7.46	9.91	17.03 ± 0.2	98
m^2Guo	3.48	5.40	8.88 ± 0.1	100
m^2_2Guo	9.91	4.04	13.65 ± 0.3	98
PCNR	8.57	4.82	13.25 ± 0.2	99
t^6Ado	4.39	2.60	6.97 ± 0.1	100

[a] Each value is an average of three assays.
[b] Standard deviation.

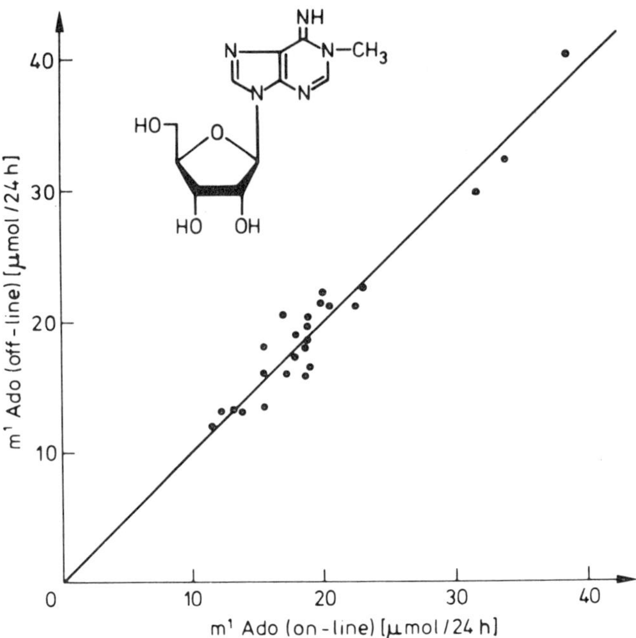

FIGURE 10. Off-line/on-line correlation. The comparison of "off-line" and "on-line"-chromatographic methods is based on the quantitative determination of m[1]Ado, m[1]Ino, m[2]Guo, and m_2^2Guo in 20 different urines and representatively shown for m[1]Ado. Each value is an average of three independent runs. Regression lines: m[1]Ado, y = 1.003 × +0.006, r = 0.9633; m[1]Ino, y = 0.894 × +2.132, r = 0.9185; m[2]Guo, y = 0.642 × +1.515, r = 0.6988; m_2^2Guo, y = 0.937 × +0.178, r = 0.9385. (Taken from Hagemeier, E., Kemper, K., Boos, K.-S., and Schlimme, E., *J. Clin. Chem. Clin. Biochem.*, 22, 175, 1984. With permission.)

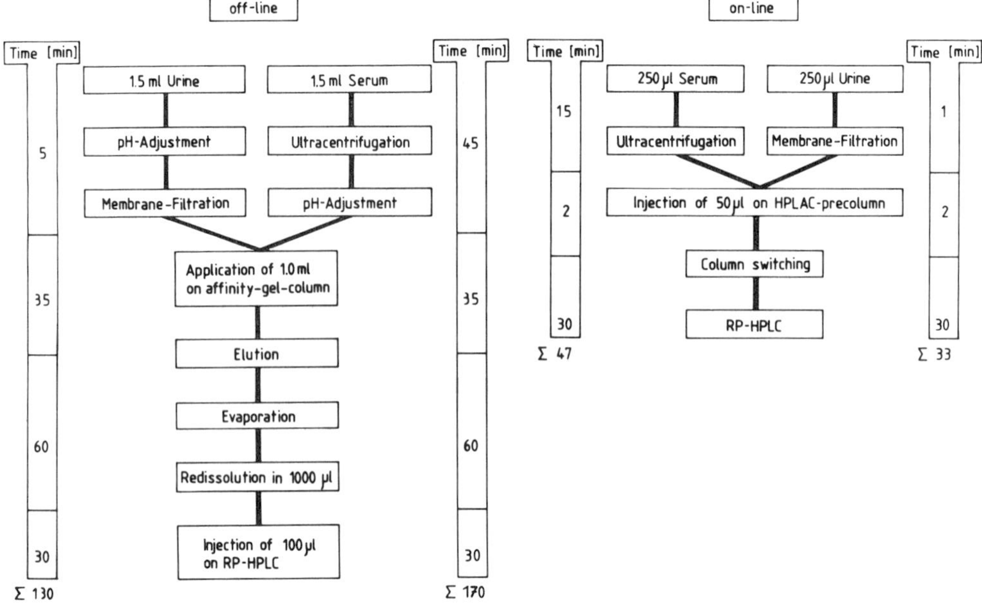

FIGURE 11. Flow diagram of analytical procedures. (Taken from Hagemeier, E., Kemper, K., Boos, K.-S., and Schlimme, E., *J. Clin. Chem. Clin. Biochem.*, 22, 175, 1984. With permission.)

Table 3
URINARY NUCLEOSIDE EXCRETION[a,b]

ψ	m^1Ado	PCNR	m^1Ino	m^1Guo	m^2Guo	m^2_2Guo	t^6Ado	Ref.
26.7 ± 4.5	1.76 ± 0.48	1.05 ± 0.26	1.18 ± 0.39	1.07	0.41 ± 0.12	1.44 ± 0.38	n.d.[c]	37
23.5	1.8	n.d.	1.2	0.6	0.3	0.8	n.d.	38
17.8 ± 3.2	4.32 ± 0.79	n.d.	1.04 ± 0.21	0.96 ± 0.22	1.29 ± 0.50	1.36 ± 0.45	n.d.	39
20.9 ± 4.3	2.15 ± 0.46	0.68 ± 0.16	1.20 ± 0.24	0.71 ± 0.16	n.d.	1.25 ± 0.30	n.d.	2
n.d.	2.1 ± 0.29	1.89 ± 0.58	1.7 ± 0.23	n.d.	0.38 ± 0.10	2.2 ± 0.29	0.73 ± 0.25	23, 34

[a] Values expressed as nmol/μmol creatinine.
[b] In case authors distinguish between female and male control groups, values are given for the female group.
[c] Not determined.

Table 4
APPLICATIONS OF BORONIC ACID-AFFINITY CHROMATOGRAPHY

Target compounds	Ref.
Catechols	40—43
Carbohydrates	25, 44
Ribonucleotides	25, 45, 46
RNA	45, 47
Glycoproteins	48—50
Phosphoribosylated proteins	51
Enzymes	52

REFERENCES

1. **Nass, G., Ed.,** *Modified Nucleosides and Cancer,* Springer-Verlag, Berlin, 1983.
2. **Heldman, D. A., Grever, M. R., Speicher, C. E., and Trewyn, R. W.,** Urinary excretion of modified nucleosides in chronic myelogenous leukemia, *J. Lab. Clin. Med.,* 101, 783, 1983.
3. **Heldman, D. A., Grever, M. R., Miser, J. S., and Trewyn, R. W.,** Relationship of urinary excretion of modified nucleosides to disease status in childhood acute lymphoblastic leukemia, *J. Natl. Cancer Inst.,* 71, 269, 1983.
4. **Fox, I. H. and Kelley, W. N.,** The role of adenosine and 2′-deoxyadenosine in mammalian cells, *Annu. Rev. Biochem.,* 47, 655, 1978.
5. **Baer, H. P. and Drummond, G. I., Eds.,** *Physiological and Regulatory Functions of Adenosine and Adenine Nucleotides,* Raven Press, New York, 1979.
6. **Sprinzl, M. and Gauss, D. H.,** Compilation of tRNA sequences, *Nucleic Acids Res.,* 12, r1, 1984.
7. **Agris, P. F., Ed.,** *Modified Nucleosides of tRNA,* 2nd ed., Alan R. Liss, New York, 1983.
8. **Chheda, G. B.,** Purine, pyrimidine, pyridine, and imidazole derivatives excreted in human urine, in *Handbook of Biochemistry and Molecular Biology: Nucleic Acids,* 3rd ed., Fasman, G., Ed., CRC Press, Boca Raton, Fla., 1975, 251.
9. **Uziel, M., Smith, L. H., and Taylor, St. A.,** Modified nucleosides in urine: selective removal and analysis, *Clin. Chem.,* 22, 1451, 1976.
10. **Davis, G. E., Suits, R. D., Kuo, K. C., Gehrke, Ch. W., Waalkes, T. P., and Borek, E.,** High-performance liquid chromatographic separation and quantitation of nucleosides in urine and some other biological fluids, *Clin. Chem.,* 23, 1427, 1977.
11. **Brown, P. R., Hartwick, R. A., and Krstulovic, A. M.,** Analysis of purine nucleosides and their bases by reverse phase partition high pressure liquid chromatography, *J. Clin. Chem. Clin. Biochem.,* 14, 282, 1976.
12. **Hartwick, R. A. and Brown, P. R.,** Evaluation of microparticle chemically bonded reversed-phase packings in the high-pressure liquid chromatographic analysis of nucleosides and their bases, *J. Chromatogr.,* 126, 679, 1976.
13. **Gehrke, Ch. W., Kuo, K. C., Davis, G. E., Suits, R. D., Waalkes, T. P., and Borek, E.,** Quantitative high-performance liquid chromatography of nucleosides in biological materials, *J. Chromatogr.,* 150, 455, 1978.
14. **Pfadenhauer, E. H. and Sun-De Tong,** Determination of inosine and adenosine in human plasma using high-performance liquid chromatography and a boronate affinity gel, *J. Chromatogr.,* 162, 585, 1979.
15. **Karle, J. M., Anderson, L. W., Dietrick, D. D., and Cysyk, R. L.,** Determination of serum and plasma uridine levels in mice, rats, and humans by high-pressure liquid chromatography, *Anal. Biochem.,* 109, 41, 1980.
16. **Schöch, G., Thomale, J., Lorenz, H., Suberg, H., and Karsten, U.,** A new method for the simultaneous analysis of unmodified and modified urinary nucleosides and nucleobases by high performance liquid chromatography, *Clin. Chim. Acta,* 108, 247, 1980.
17. **Schlimme, E., Boos, K.-S., and Weise, M.,** Selektive Charakterisierung von N1-Methyladenosin neben N7-Methylguanosin im Harn, *J. Clin. Chem. Clin. Biochem.,* 19, 55, 1981.
18. **Colonna, A., Russo, T., Esposito, F., Salvatore, F., and Cimino, F.,** Determination of pseudouridine and other nucleosides in human blood serum by high-performance liquid chromatography, *Anal. Biochem.,* 130, 19, 1983.
19. **Huguenin, P. N., Jayaram, H. N., and Kelley, J. A.,** Reverse phase HPLC determination of 5,6-dihydro-5-azacytidine in biological fluids, *J. Liq. Chromatogr.,* 7, 1433, 1984.
20. **Freeman, D. H.,** Review: ultraselectivity through column switching and mode sequencing in liquid chromatography, *Anal. Chem.,* 53, 2, 1981.
21. **Apffel, J. A., Alfredson, T. V., and Majors, R. E.,** Automated on-line multi-dimensional high-performance liquid chromatographic techniques for the clean-up and analysis of water-soluble samples, *J. Chromatogr.,* 206, 43, 1981.
22. **Hagemeier, E., Boos, K.-S., Schlimme, E., Lechtenbörger, K., and Kettrup, A.,** Synthesis and application of a boronic acid-substituted silica for high-performance liquid affinity chromatography, *J. Chromatogr.,* 268, 291, 1983.
23. **Hagemeier, E.,** *Entwicklung eines "On-line"-Chromatographischen HPLAC-HPLC-Verfahrens zur Quantitativen Bestimmung von Ribonucleosiden in Körperflüssigkeiten,* dissertation, Universität Paderborn, Paderborn, 1983.
24. **Ohlson, S., Hansson, L., Larsson, P.-O., and Mosbach, K.,** High performance liquid affinity chromatography (HPLAC) and its application to the separation of enzymes and antigens, *FEBS Lett.,* 93, 5, 1978.

25. **Glad, M., Ohlson, S., Hansson, L., Mansson, M.-O., and Mosbach, K.,** High-performance liquid affinity chromatography of nucleosides, nucleotides and carbohydrates with boronic acid substituted microparticulate silica, *J. Chromatogr.*, 200, 254, 1980.
26. **Larsson, P.-O., Glad, M., Hansson, L., Mansson, M.-O., and Mosbach, K.,** High-performance liquid affinity chromatography, *Adv. Chromatogr.*, 21, 41, 1983.
27. **Hagemeier, E., Kemper, K., Boos, K.-S., and Schlimme, E.,** On-line high-performance liquid affinity chromatography — high-performance liquid chromatography analysis of monomeric ribonucleoside compounds in biological fluids, *J. Chromatogr.*, 282, 663, 1983.
28. **Hagemeier, E., Kemper, K., Boos, K.-S., and Schlimme, E.,** Characterization and quantitative determination of minor ribonucleosides in physiological fluids. I. Development of a chromatographic method for the quantitative determination of minor ribonucleosides in physiological fluids, *J. Clin. Chem. Clin. Biochem.*, 22, 175, 1984.
29. **Hartwick, R. A., Assenza, S. P., and Brown, P. R.,** Identification and quantitation of nucleosides, bases, and other UV-absorbing compounds in serum, using reversed-phase high-performance liquid chromatography. I. Chromatographic methodology, *J. Chromatogr.*, 186, 647, 1979.
30. **Gehrke, Ch. W., Kuo, K. C., and Zumwalt, R. W.,** Chromatography of nucleosides, *J. Chromatogr.*, 188, 129, 1980.
31. **Assenza, S. P. and Brown, P. R.,** Evaluation of reversed-phase, radially compressed, flexible-walled columns for the separation of low molecular weight, UV-absorbing compounds in serum, *J. Liq. Chromatogr.*, 3, 41, 1980.
32. **Ramsteiner, K. A. and Böhm, K. H.,** Automatic device for injection and multiple column switching in high-performance liquid chromatography, *J. Chromatogr.*, 260, 33, 1983.
33. **Little, C. J., Tompkins, D. J., Stahel, O., Frei, R. W., and Werkhoven-Goewie, C. E.,** Applications of a microprocessor-controlled valve-switching unit for automated sample cleanup and trace enrichment in high-performance liquid chromatography, *J. Chromatogr.*, 264, 183, 1983.
34. **Kemper, K.,** *Multidimensionale Hochleistungsflüssigchromatographie von Biomolekülen mit Cis-Diol-Funktion*, dissertation, Universität Paderborn, Paderborn, 1985.
35. **Hartwick, R. A., Krstulovic, A. M., and Brown, P. R.,** Identification and quantitation of nucleosides, bases and other UV-absorbing compounds in serum, using reversed-phase high-performance liquid chromatography. II. Evaluation of human sera, *J. Chromatogr.*, 186, 659, 1979.
36. **Ontyd, J. and Schrader, J.,** Measurement of adenosine, inosine, and hypoxanthine in human plasma, *J. Chromatogr.*, 307, 404, 1984.
37. **Borek, E., Sharma, O. K., and Waalkes, T. P.,** New applications of urinary nucleoside markers, in *Modified Nucleosides and Cancer*, Nass, G., Ed., Springer-Verlag, Berlin, 1983, 301.
38. **Rasmuson, T., Björk, G. R., Damber, L., Holm, S. E., Jacobsson, L., Jeppsson, A., Littbrand, B., Stigbrand, T., and Westman, G.,** Evaluation of carcinoembryonic antigen, tissue polypeptide antigen, placental alkaline phosphatase, and modified nucleosides as biological markers in malignant lymphomas, in *Modified Nucleosides and Cancer*, Nass, G., Ed., Springer-Verlag, Berlin, 1983, 331.
39. **Salvatore, F., Colonna, A., Costanzo, F., Russo, T., Esposito, F., and Cimino, F.,** Modified nucleosides in body fluids of tumor-bearing patients, in *Modified Nucleosides and Cancer*, Nass, G., Ed., Springer-Verlag, Berlin, 1983, 360.
40. **Higa, S., Suzuki, T., Hayaski, A., Tsuge, I., and Yamamura, Y.,** Isolation of catecholamines in biological fluids by boric acid gel, *Anal. Biochem.*, 77, 18, 1977.
41. **Speek, A. J., Odink, J., Schrijver, J., and Schreurs, W. H. P.,** High-performance liquid chromatographic determination of urinary free catecholamines with electrochemical detection after pre-purification on immobilized boric acid, *Clin. Chim. Acta*, 128, 103, 1983.
42. **Hansson, L., Glad, M., and Hansson, C.,** Boronic acid-silica: a new tool for the purification of catecholic compounds on-line with reversed-phase high-performance liquid chromatography, *J. Chromatogr.*, 265, 37, 1983.
43. **Boos, K.-S., Wilmers, B., Sauerbrey, R., and Schlimme, E.,** Development of a fully automatic catecholamine analyser, *Fresenius Z. Anal. Chem.*, 324, 320, 1986.
44. **Weith, H. L., Wiebers, J. L., and Gilham, P. T.,** Synthesis of cellulose derivatives containing the dihydroxyboryl group and a study of their capacity to form specific complexes with sugars and nucleic acid components, *Biochemistry*, 9, 4396, 1970.
45. **Rosenberg, M., Wiebers, J. L., and Gilham, P. T.,** Studies on the interactions of nucleotides, polynucleotides and nucleic acids with dihydroxyboryl-substituted celluloses, *Biochemistry*, 11, 3623, 1972.
46. **Moore, E. C., Peterson, D., Li Ying Yang, Chou Yau Yeung, and Faye Neff, N.,** Separation of ribonucleotides and deoxyribonucleotides on columns of borate covalently linked to cellulose. Application to the assay of ribonucleoside diphosphate reductase, *Biochemistry*, 13, 2904, 1974.
47. **Singhal, R. P., Bajaj, R. K., Buess, C. M., Smoll, D. B., and Vakharia, V. N.,** Reversed-phase boronate chromatography for the separation of O-methylribose nucleosides and aminoacyl-tRNAs, *Anal. Biochem.*, 109, 1, 1980.

48. **Mallia, A. K., Hermanson, G. T., Krohn, R. I., Fujimoto, E. K., and Smith, P. K.,** Preparation and use of a boronic acid affinity support for separation and quantitation of glycosylated hemoglobins, *Anal. Lett.*, 14 (B8), 649, 1981.
49. **Bruns, D. E., Lobo, P. I., Savery, J., and Wills, R.,** Specific affinity-chromatographic measurement of glycated hemoglobins in uremic patients, *Clin. Chem.*, 30, 569, 1984.
50. **Yatscoff, R. W., Tevaarwerk, G. J. M., Clarson, C. L., and Warnock, L. M.,** Evaluation of an affinity chromatographic procedure for the determination of glycosylated hemoglobin (HbA_1), *Clin. Biochem.*, 16, 291, 1983.
51. **Romaschin, A. D., Kirsten, E., Jackowski, G., and Kun, E.,** Quantitative isolation of oligo- and polyadenosinediphosphorylated proteins by affinity chromatography from livers of normal and dimethylnitrosamine-treated Syrian hamsters, *J. Biol. Chem.*, 256, 7800, 1981.
52. **Bouriotis, V., Galpin, F. J., and Dean, P. D. G.,** Applications of immobilised phenylboronic acids as supports for group-specific ligands in the affinity chromatography of enzymes, *J. Chromatogr.*, 21, 267, 1981.
53. **Schlimme, E., Boos, K.-S., Hagemeier, E., Kemper, K., Meyer, U., Hobler, H., Schnelle, T., and Weise, M.,** Direct clean-up and analysis of ribonucleosides in physiological fluids, *J. Chromatogr.*, 378, 349, 1986.

LABELING OF PYRIMIDINE COMPOUNDS FOR FLUORESCENT DETECTION IN HIGH-PERFORMANCE LIQUID CHROMATOGRAPHY

Shigeru Yoshida, Shingo Hirose, and Masaki Iwamoto

INTRODUCTION

Fluorescence labeling of nonfluorescent compounds facilitates their determination by high-performance liquid chromatography (HPLC) due to the inherently higher detection sensitivity compared to UV absorption. In addition, the linearity of response is expanded over several orders of magnitude.

Fluorescent derivatization of purine compounds[1-3] has been used in combination with HPLC in nucleic acid research. However, the derivatization of pyrimidine nucleobases and nucleosides has not been studied extensively.

4-Bromomethyl-7-methoxycoumarin (Br-Mmc)[4-6] has been shown to form fluorescent esters with monocarboxylic acids. Reactions of Br-Mmc with other OH, NH_2, and NHR groups have been investigated in aqueous media, but with little success.

Using this reagent, Dünges and Seiler[7] derivatized in a nonaqueous medium some barbiturates which are imide compounds, as well as pyrimidine nucelobases and nucleosides. Thus, they demonstrated the usefulness of Br-Mmc for fluorescence labeling of pyrimidine compounds.

The aim of this section is to demonstrate the applicability and scope of Br-Mmc labeling in trace analysis of pyrimidine nucleobases, nucleosides, and related compounds. A simple and virtually quantitative esterification method yields stable derivatives that are amenable to reversed-phase HPLC analysis.

ESTERIFICATION WITH Br-Mmc

Labeling Method

Use of Crown Ether Catalysts

A tenfold excess of Br-Mmc, a twofold excess of 18-crown-6 ether, and 25 mg of crystalline water-free potassium carbonate were added to 0.5 mg of pyrimidine compounds (total weight) in 20 mℓ of acetone-acetonitrile mixture (volume ratio, 1:2) (protected from light by wrapping the flask in aluminum foil). The mixture was allowed to reflux for 45 min in a water bath. After cooling, 0.2 mℓ of *n*-valeric acid was added to the mixture for the esterification of excess Br-Mmc. The mixture was refluxed again for 5 min. About 10 μℓ of this solution was applied to a thin-layer chromatography (TLC) plate. For the HPLC analysis, the mixture was diluted 10^2- to 10^5-fold with acetone.

Utilization of Solvation Effects

Mmc derivatives were prepared by adding Br-Mmc (10 mg) to a solution containing pyrimidine compounds (total weight, within 1 mg) and potassium carbonate (10 mg) in 5 mℓ of dimcthylsulfoxide (DMSO). The reaction was completed within 5 min at room temperature. The excess Br-Mmc was treated with *p*-nitrobenzoic acid which gave a non-fluorescent Mmc derivative.

Treatment of Excess Br-Mmc

A critical part of this labeling method is the interfering peak of unreacted Br-Mmc, which can be eliminated by derivatizing excess Br-Mmc after the labeling reaction. For easy esterification of excess Br-Mmc, *n*-valeric acid, hordenine, and *p*-nitrobenzoic acid can be used; *p*-nitrobenzoic acid gives an Mmc derivative which does not fluoresce (Figure 1).

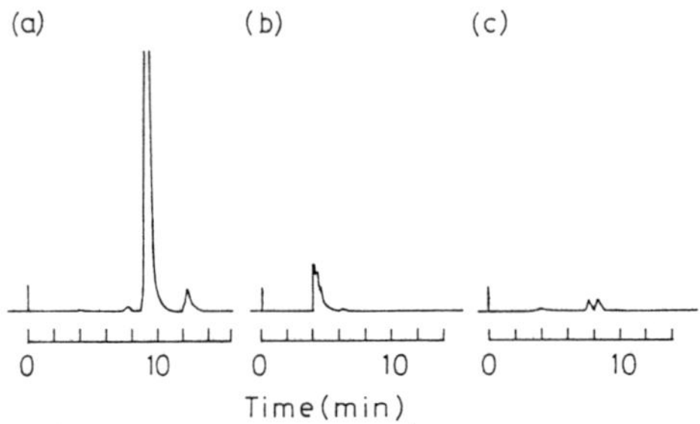

FIGURE 1. Treatment of excess Br-Mmc. Column: Nucleosil® 5, C_{18} (Machery, Nagel & Co., Düren, West Germany); 20 cm × 4 mm ID. Mobile phase: 80% (v/v) MeOH in water; flow rate: 0.6 mℓ/min. (a) With *n*-valeric acid; (b) with hordenine; (c) with *p*-nitrobenzoic acid.

Reaction Mechanisms

Structure of the Derivatives

The structure of the alkylated derivatives has been confirmed by many workers.[8-11] The possible labeling reaction of pyrimidine nucleobases and nucleosides are shown in Scheme I. The separation of *N*-mono-Mmc and *N,N'*-bis-Mmc derivatives of 5-fluorouracil (5FU) is shown in Figure 2.

CH_3O … CH_2Br

(Br - Mmc)

Br - Mmc, refluxing acetone, K_2CO_3, 18-CROWN-6

Br - Mmc, K_2CO_3 DMSO

Br - Mmc

Br - Mmc

The fluorescence intensities of the mono- and the bis-derivative differ markedly; the quantum yield of the bis-derivative is higher than that of the mono-derivative. On the other hand, the chromatogram of Mmc-ftorafur (FT, 1-[tetrahydro-2-furanyl]-5-fluorouracil) derivative does not contain other undesirable components.

These labeling reactions were also studied in detail in conjunction with TLC. With the

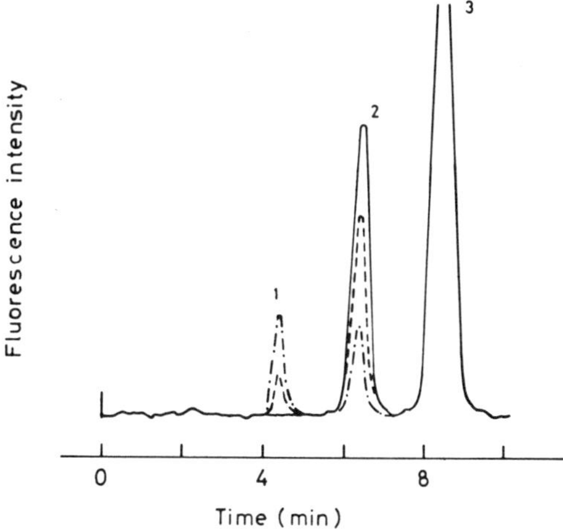

FIGURE 2. Separation and formation process of *N*-mono-Mmc and *N,N'*-bis-Mmc derivatives of 5FU. Column: Nucleosil® 5, C_{18} (Machery, Nagel & Co., Düren, West Germany); 20 cm × 4 mm ID. [Br-Mmc]/[5FU] = 10. (1) *N*-mono-Mmc derivative of 5FU; (2) *N,N'*-bis-Mmc derivative of 5FU; (3) Mmc derivative of *n*-valeric acid. -·-·-: After 15 min of reaction initiation; - - - - -: after 30 min; ———: after 45 min. After each reaction time, 0.2 mℓ of *n*-valeric acid was added to the reaction mixture and the solution was refluxed for 5 min at 75°C. Mobile phase: 70% (v/v) methanol in water; flow rate: 0.8 mℓ/min.

solvent system of ethyl acetate-methanol-10% formic acid (9:1:1, v/v) a satisfactory separation was obtained with R_f values of 0.89 for the *N,N'*-bis-Mmc derivative of 5FU, 0.87 for the Mmc derivative of FT, and 0.65 for the *N*-mono-derivative of 5FU. The Mmc derivative of *n*-valeric acid (esterification of excess Br-Mmc) had an R_f value of 0.93. The spot of the *N*-mono-Mmc derivative of 5FU gradually disappeared in the course of the reaction and was not seen 45 min after the start of the reaction. Other fluorescent spots, which could represent by-products of Br-Mmc and other undesirable components, were well separated in both systems.

The spots of the Mmc derivatives on the silica gel plates were stable for several days without spraying or other special protective procedures. By repeated chromatography of the same sample solution it was shown that the Mmc derivatives were stable for at least several weeks in the reaction mixtures.

Crown Ether Catalysts and Solvation Effects

Crown ethers have been shown to have a tremendous ability to complex metal salts,[12] especially those of potassium. Anions of these salts in solution have been shown to be unusually reactive, especially the carboxylate anions.[6,13] One of the most interesting and important properties of solid-liquid crown ether phase transfer of salts is that stoichiometric concentrations of the crown ethers are not necessary.[14] Thus, one may use crown ethers in molar ratios of 1:20 to 1:100 to catalyze the phase transfer of carboxylate salts. Under these conditions, the nucleophilic properties of the carboxylate anion are enhanced and 92 ~ 98% yields have been reported.[15] Another example of the application of this method has been reported by Dünges and Seiler.[7] A barbiturate ester of Br-Mmc was derivatized by refluxing acetone solution of the reaction components in the presence of crystalline water-free K_2CO_3 with crown ethers as catalysts. The barbiturates usually contain two acidic imide groups,

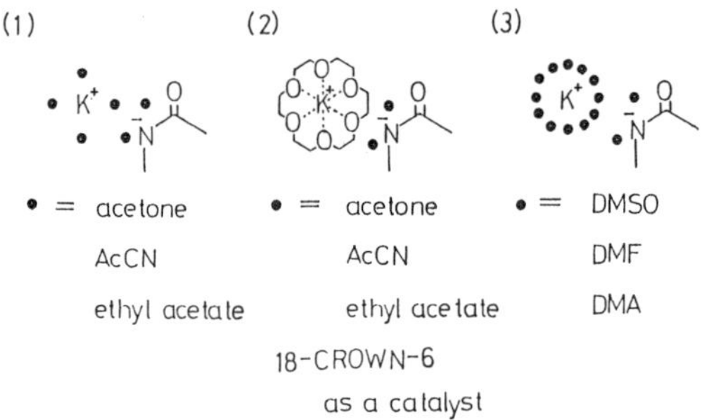

FIGURE 3. Effect of solvated ion.

each of which can be alkylated at the N atom. The pyrimidine nucleobases and nucleosides, as well as the barbiturates, can also be alkylated by alkyl halide, and the crown ether-potassium complex provides increased anion reactivity, as shown in Figure 3(2). The crown ether (18-crown-6) is used as catalyst because it is inexpensive.

On the other hand, small ''closed-shell'' cations like Na^+ and K^+ are more solvated by dimethylacetoamide (DMA), dimethylformamide (DMF), and DMSO than by acetone.[16] A number of investigators have used a solution of Na^+ and K^+ in dry DMSO in order to permethylate acidic compounds. The strongly basic methylsulfinyl carbanion which is formed in the solvent deprotonates even very weak acids, which are then able to react with methyl iodide. The reaction is fast and quantitative at room temperature. This procedure has enabled successful GLC determinations of bile acids,[17] nucleotides,[18] 5FU,[9] and 5-fluorouridine.[19] When Br-Mmc is used instead of methyliodide, same solvation effects are expected to label pyrimidine bases and nucleosides in the solvent containing Na^+ and K^+, as shown in Figure 3(3). The labeling reaction is fast at room temperature.

Fluorescence Properties of the Derivatives

Fluorescence Quantum Yields

Quantum yields of fluorescence are estimated by the comparative method of Parker and Ree[20] using tryptophan (20%)[21] as the reference of known quantum yield.

$$Q = Q_0 \times A_0/F_0 \times F/A \qquad (1)$$

where Q and Q_0 denote the quantum yield of the unknown and of the reference (tryptophan), respectively; A and A_0 are absorbancies of the unknown and reference solution used for fluorescence assay; F and F_0 denote the areas under the fluorescence spectra of the unknown and reference solution.

UV absorbance and fluorescence spectral data and quantum yields for some esters of pyrimidine nucleobases and nucleosides and Br-Mmc in 70% (v/v) methanol are given in Table 1.

The uncorrected fluorescence excitation and emission spectra of the Mmc-FT derivative in 70% (v/v) methanol are shown in Figure 4.

Spectral Characteristics and Solvent Effects

The effect of the variation of methanol concentration in water on fluorescence intensity excited at 333 nm and monitored at 396 nm is shown in Figure 5. The derivatives used are listed in Table 1. Upon addition of methanol to the aqueous solution of the derivatives, an initial rapid increase of fluorescence intensity of pyrimidine nucleobases is observed com-

Table 1
UV ABSORPTION AND FLUORESCENCE DATA FOR SOME PYRIMIDINE NUCLEOBASES AND NUCLEOSIDES IN 70% (v/v) METHANOL IN WATER

	Ex max (nm)	Em max (nm)	Q	UV max (nm)	ϵ(cm$^{-1}M^{-1}$) × 10^2
Mmc-Ura	333	400	0.13		
Mmc-5FU	333	396	0.14		
Mmc-Thy	333	397	0.14		
Mmc-Urd	330	396	0.11	329	127
Mmc-dUrd	329	398	0.11	329	125
Mmc-FdUrd	334	397	0.12	335	135
Mmc-Tdr	329	396	0.10	328	129
Mmc-Ino	330	399	0.10	329	129
Br-Mmc	325	396	0.02	326	120

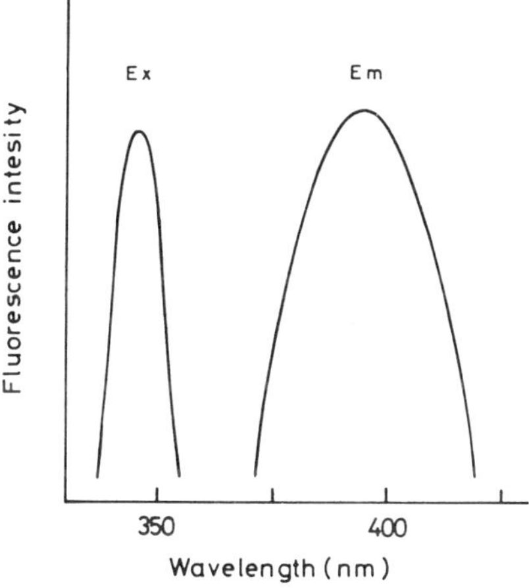

FIGURE 4. Fluorescence spectra of FT-Mmc derivative (uncorrected for solvent absorption). Ex: excitation; Em: emission. Solvent: 70% (v/v) methanol in water.

pared with the nucleosides. The maximum fluorescence intensity for the nucleobase derivatives is at a methanol content of 70% (v/v). Above this level, a slow decrease of intensity occurs. The results are similar for other nucleosides. It is well known that the fluorescence of many compounds depends on the water content of the solvent.[22] Thus, the HPLC separation employing mobile phases of higher water content will not give high sensitivity typical of fluorescence detection. However, reversed-phase separations of higher molecular weight derivatives can be performed using mobile phases containing approximately 70% (v/v) of methanol. This affords high fluorescence intensities.

TLC Separation

Separations were carried out after spotting with 10 $\mu\ell$ micropipettes (Drummond Scientific Company). Two solvent systems ([1]ethylacetate and [2]ethylacetate-methanol-10% [v/v] formic acid in water [volume ratio, 9:1:1]) were used for the separation on a silica gel 60, F-254 plate (Merck). The plate was developed in a paper-lined chromatography chamber over a distance of 15 cm, removed, and air dried. The plate was examined under UV light

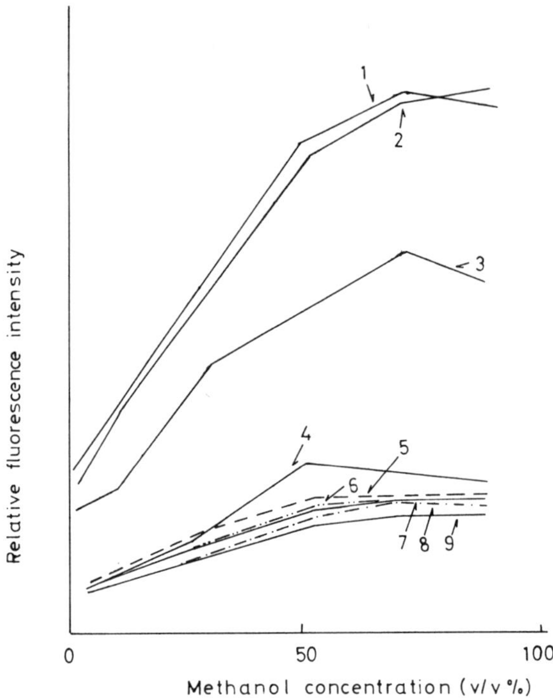

FIGURE 5. Variation of fluorescent intensity effected with MeOH concentration. (1) Mmc-5FU; (2) Mmc-Thy; (3) Mmc-Ura; (4) Mmc-FdUrd; (5) Mmc-Ino; (6) Mmc-Urd; (7) Mmc-dUrd; (8) Mmc-Tdr; (9) Mmc-Ade.

at 254 and 365 nm. R_f values of pyrimidine derivatives and their related compounds are listed in Table 2.

As the impurities of Br-Mmc hindered the identification of Mmc esters in HPLC, the TLC spot of interest was scraped, transferred into a centrifuge tube, and extracted by shaking for 5 min in 5 mℓ of acetone. The centrifuged solution was evaporated to dryness under vacuum; 100 μℓ of acetone was added to the residue and the solution was analyzed by HPLC.

HPLC Analysis

The analysis was performed using a Shimadzu (Kyoto, Japan) Model LC-4A chromatograph equipped with a Model SIL-1A injector. Chromatographic separations were performed using a Nucleosil® 5, C_{18} column (20 cm × 4 mm I.D., 5 μm) (Machery, Nagel & Co., Düren, West Germany). The column was packed using a balanced density slurry procedure similar to that described by Majors.[23] The mobile phase was filtered and vacuum degassed before use. The flow rate was 0.6 mℓ/min and the temperature was ambient.

The column effluent was monitored by fluorimetrically at the excitation and emission wavelengths of 346 and 395 nm, respectively, using a Shimadzu Model RF-530 fluorescence spectrophotometer.

SELECTED APPLICATIONS

Pyrimidine Nucleobases

Figure 6 shows the HPLC separation of Mmc derivatives of uracyl (Ura), 5FU, and thymine (Thy) using a reversed-phase column and the mobile phase consisting of CH_3OH-

Table 2
TLC SEPARATION OF Mmc-PYRIMIDINE DERIVATIVES, RIBOSE, AND DEOXYRIBOSE

	R_f value	
	Solvent I[a]	Solvent II[b]
Mmc-Ura	0.74, 0.58	0.48, 0.30
Mmc-Thy	0.81, 0.68	0.57, 0.39
Mmc-5FU	0.89	0.74
Mmc-Urd	0.30	0.03
Mmc-dUrd	0.40	0.12
Mmc-FdUrd	0.57	0.21
Mmc-Thy	0.44	0.08
Ribose	—	—
Deoxyribose	—	—

[a] Solvent I: ethyl acetate-MeOH-10% formic acid (9:1:1 v/v).
[b] Solvent II: ethyl acetate.

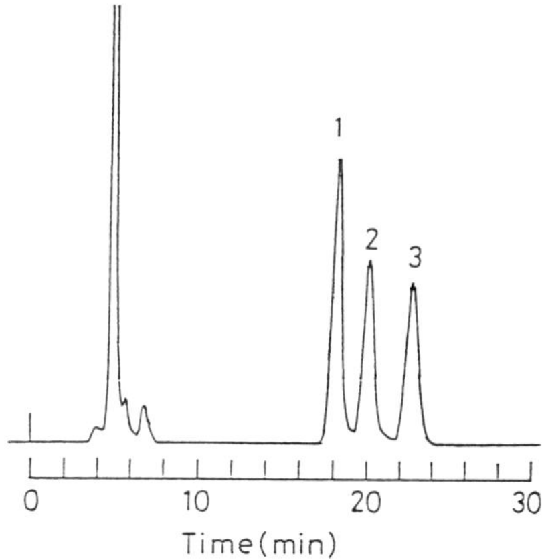

FIGURE 6. Separation of Mmc-pyrimidine nucleobases. Column: Nucleosil® 5, C_{18} (Machery, Nagel & Co., Düren, West Germany); 20 cm × 4 mm ID. Mobile phase: CH_3OH-CH_3CN-water = 45:10:45 (v/v); flow rate: 0.8 mℓ/min. (1) Mmc-Ura; (2) Mmc-5FU; (3) Mmc-Thy.

CH_3CN-water (45:10:45 v/v) at a flow rate of 0.8 mℓ/min. The excess Br-Mmc was treated with *p*-nitrobenzoic acid. Typical detection limits were 5 pg for Ura, 15 pg for Thy, and 10 pg for 5FU. Further optimization of analytical conditions will enable the detection of subpicogram levels.

Pyrimidine Nucleosides and 5FU

Figure 7 shows the separation of the Mmc derivatives of four pyrimidine nucleosides (uridine [Urd], deoxyuridine [dUrd], fluorodeoxyuridine [FdUrd], and thymidine [Tdr]) and 5FU. After approximately 25 min, the mobile phase is changed from 54% (v/v) methanol to 65% methanol by a stepwise gradient, and after 35 min it is changed to 100% methanol

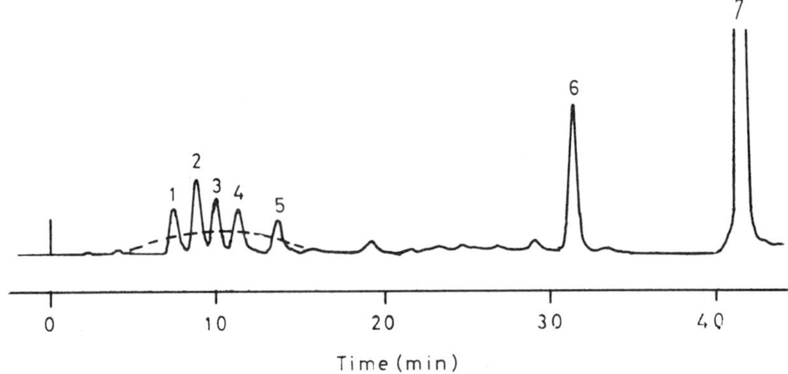

FIGURE 7. Separation of Mmc-pyrimidine and Mmc-Urd derivatives. Column: Nucleosil® 5, C_{18} (Machery, Nagel & Co., Düren, West Germany); 20 cm × 4 mm ID. - - - - -: ψUrd. (1) Urd; (2) impurity; (3) dUrd; (4) FdUrd; (5) Tdr; (6) 5FU; (7) *n*-valeric acid. Mobile phase: from 0 to 20 min — 54% methanol in water; from 20 to 35 min — 65% methanol in water; from 35 to 40 min — 100% methanol. Flow rate: 0.6 mℓ/min.

(these derivatives are labeled by refluxing in acetone). The impurity peak in Figure 7 formed when the labeling reaction is carried out by refluxing in acetone can interfere in the determination of some nucleoside derivatives (dUrd and Tdr).

The limits of detection are an important criterion in trace work. Using this derivatization method of the Tdr derivative 75 pg can be detected. The detection limits for 5FU, Urd, dUrd, and FdUrd derivative are 2.5, 5.0, 50, and 50 pg, respectively. These detection limits can be improved by further optimization of the efficiency of chromatographic separation.

Pseudouridine and 5FU

Pseudouridine (ψUrd), one of the modified nucleosides derived from the enzymatic degradation of RNA (particularly the transfer RNA), has been detected in human urine.[24] The urinary excretion of modified nucleosides, and particularly ψUrd, has been found to increase in certain neoplastic diseases.[25,26] These results suggest that ψUrd may be used as a tumor marker to follow the progress of the neoplastic disease and to evaluate the response to the therapy.

Simultaneous determination of 5FU and ψUrd is particularly interesting. Because of the low levels of nucleosides in biological samples and the complexity of matrices, their determinations require the use of sensitive and specific analytical methods. Although Mmc-pyrimidine, Mmc-nucleoside, and Mmc-5FU derivatives are successfully separated by a step-wise gradient with 54% (v/v) methanol and with 65% methanol, the Mmc-ψUrd derivative peak is not fully resolved (broken line in Figure 7). Therefore, an alternative method for the analysis of both Mmc-ψUrd and Mmc-5FU derivatives was developed. This was achieved by isocratic elution using a 70% (v/v) methanol solution which enabled the separation of Mmc-ψUrd and Mmc-5FU derivatives from the other nucleoside derivatives (Figure 8). However, preliminary purification of the mixture of derivatives by TLC on a silica gel plate may be useful for the analysis of blood extract, because many interfering constituents can thus be removed. The detection limits were 77 fmol (corresponding to 1 μℓ injection) for 5FU and 410 fmol for ψUrd (at a signal-to-noise ratio of 2:1). Since the ψUrd concentration in human serum is about 2 to 4 nmol/mℓ, the reported operative range shown in Figure 9 illustrates the suitability of this method for measurements of high concentrations of ψUrd in serum. In addition, ψUrd can be detected down to 410 fmol/μℓ injected.

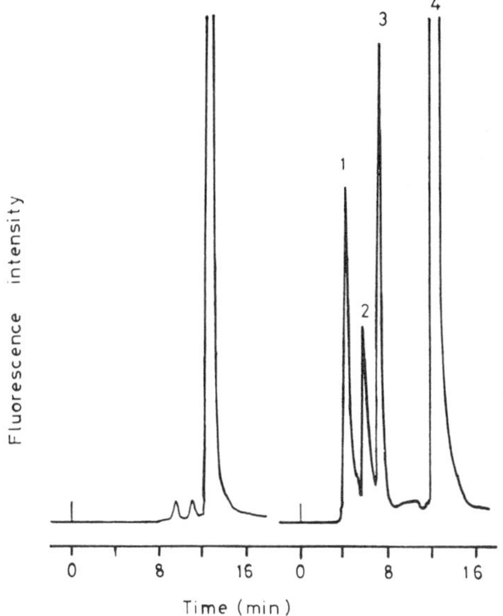

FIGURE 8. Separation of Mmc-ψUrd and Mmc-5FU derivatives. (Left) Reagent blank extracted from the TLC plate; (Right) Mmc derivatives of authentic samples extracted from the TLC plate. (1) Nucleosides; (2) ψUrd (2.2 ng); (3) 5FU(540 pg); (4) *n*-valeric acid. Column: Nucleosil® 5, C_{18} (Machery, Nagel & Co., Düren, West Germany); 20 cm × 4 mm ID. Mobile phase: 70% methanol in water; flow rate: 0.6 mℓ/min.

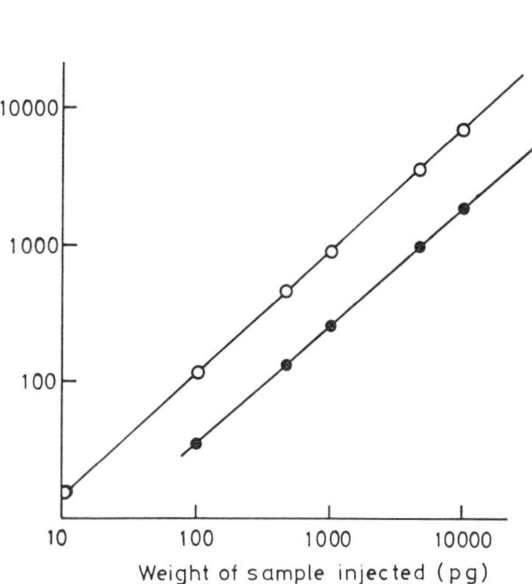

FIGURE 9. Relationship between fluorescence intensity and amount of sample. ○——○: Mmc-5FU derivative; ●——●: Mmc-ψUrd derivative.

Determination of 5FU and FT in Blood Serum

Preparation of Derivatives for Serum Analysis

Serum samples (0.5 mℓ) were adjusted with physiological saline to a total volume of 1.0 mℓ, and 0.1 mℓ of 0.5 *M* NaH_2PO_4 buffer and 8 mℓ of ethylacetate were added. After extraction and centrifugation, the organic layer was removed[27] and evaporated under vacuum. To the residue, 1 mℓ of acetone-acetonitrile mixture containing 0.5 mg/mℓ of Br-Mmc, 0.1 mg/mℓ of 18-crown-6, and 1 mg of K_2CO_3 was added and the mixture was refluxed. The resulting mixture of Mmc esters was utilized directly for HPLC analysis.

Separation by HPLC

Figure 10B shows a chromatogram of the Mmc-5FU and Mmc-FT derivatives in a serum sample and Figure 10A is a chromatogram of a serum extract free from 5FU and FT (blank sample). The blank sample shows that the serum constituents do not interfere with the separation of the 5FU and FT derivatives.

Peak identities were assigned on the basis of cochromatography with reference derivatives and comparison of retention times. The blank peaks in Figure 10 increased when using Br-Mmc stock solution which was exposed to light. When the derivatizing reagent was protected from light, the resulting Mmc-5FU and Mmc-FT derivatives were almost completely separated from the blank peaks, as can be seen in Figures 10A and 10B. Thus, it is important that the derivatization reaction be carried out away from light.

Calibration Curves and Detection Limit

To be useful in quantitative analysis, the amount of ester formed in the derivatization reaction should be related to the amount of FT and 5FU. Therefore, serum samples were spiked with increasing amounts of FT and 5FU (final concentrations of 0.02, 0.05, 0.1 to

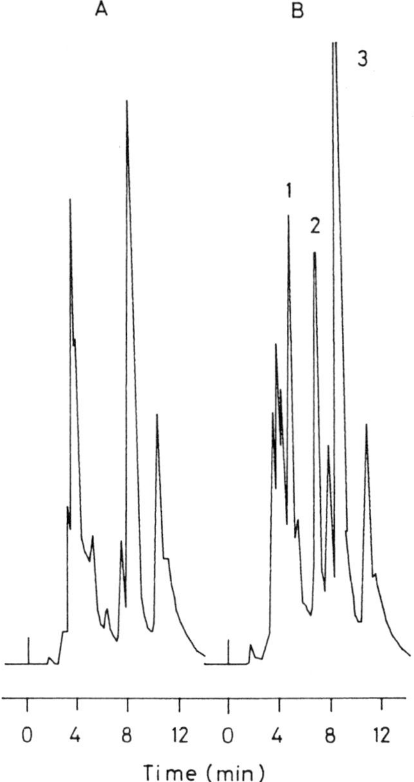

FIGURE 10. Separation of Mmc-5FU and Mmc-FT derivatives in a serum sample. (A) Blank serum sample; (B) serum sample spiked with 10 μg of 5FU and FT per serum mililiter. Column: Nucleosil® 5, C_{18}; (Machery, Nagel & Co., Düren, West Germany); 20 cm × 4 mm ID. Mobile phase: 70% (v/v) methanol in water; flow rate: 0.8 mℓ/min. (1) Mmc-FT derivative; (2) Mmc-5FU derivative; (3) Mmc-*n*-valeric derivative.

10 μg of each substance per milliliter of serum). The samples were carried through the described extraction procedure and standard curves were generated for each series by plotting peak height against known drug concentrations (Figure 11). Figure 11 illustrates the linearity of calibration curves used for quantification of FT and 5FU in serum. The overall recoveries including the extraction, Mmc derivatization, and quantification were 85 ± 5% for FT and 55 ± 5% for 5FU. Under the described separation conditions, typical detection limits were 384 fmol for FT and 100 fmol for 5FU.

Since the derivatization reaction can be scaled down to a volume of 10 μℓ,[28] it is possible to determine a few femtomoles of the derivatized compounds.

Determination of FdUrd in Serum

The internal standard for the assay of FdUrd was chlorodeoxyuridine (CldUrd) (1.0 μg/100 μℓ). The extraction and derivatization procedures are summarized in Figure 12.

Fluorescence labeling of FdUrd was done using Br-Mmc. The reaction mixture was first purified by TLC, and the known spots were isolated, extracted, and subsequently analyzed by HPLC. The HPL chromatogram is shown in Figure 13.

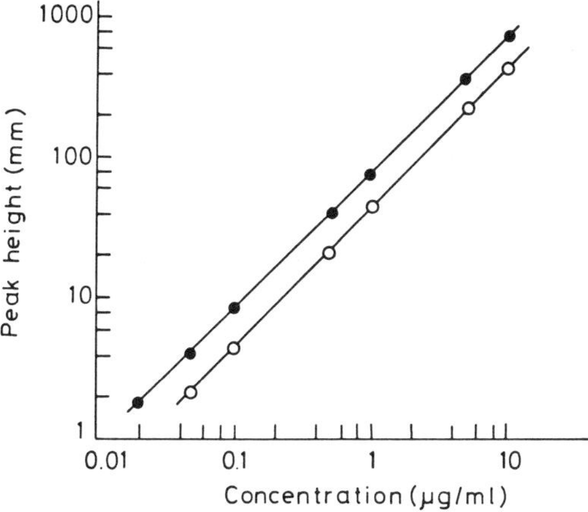

FIGURE 11. Calibration curves of Mmc derivatized 5FU and FT extracted from serum. ●——●: 5FU, ○——○: FT.

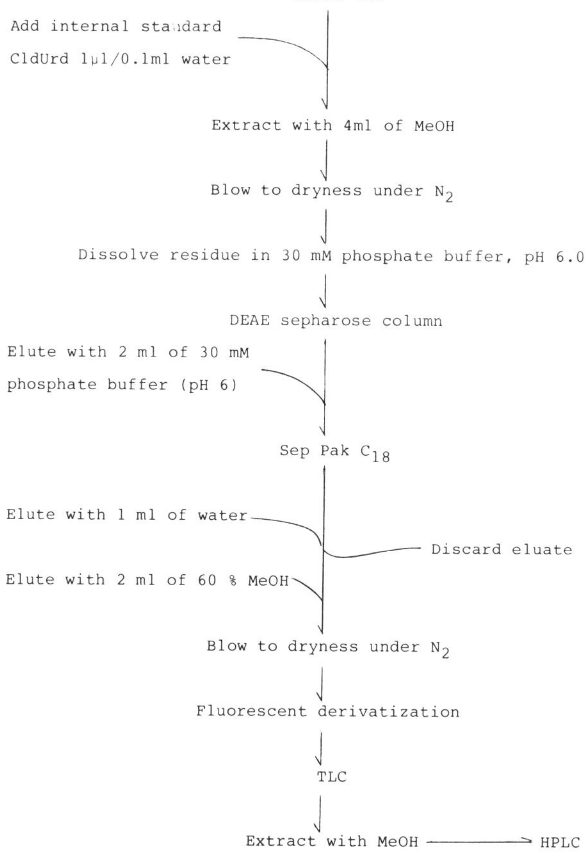

FIGURE 12. Schematic diagram of the analytical procedure for the quantitative determination of FdUrd in serum.

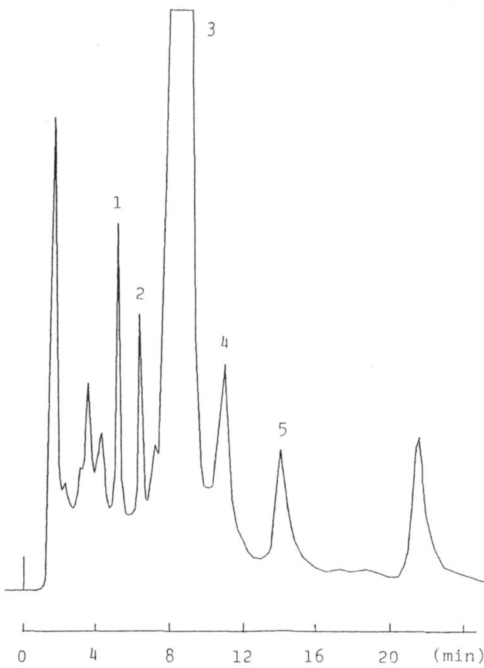

FIGURE 13. Chromatogram of a serum sample spiked with FdUrd. Column: Nucleosil® 5, C_{18} (Machery, Nagel & Co., Düren, G.F.R.); 20 cm × 4 mm ID. Mobile phase: MeOH-CH_3CN-water = 29:5:66 (v/v); flow rate: 1.0 mℓ/min; detection: Ex — 330 nm, Em — 400 nm. (1) Mmc-Ino; (2) Mmc-Urd; (3) unknown; (4) Mmc-FdUrd; (5) Mmc-CldUrd (internal standard).

REFERENCES

1. **Slowiaczek, P. and Tattersall, M. H. N.,** The determination of purine levels in human and mouse plasma, *Anal. Biochem.*, 125, 6, 1982.
2. **Secrist, J. A., Barrio, J. R., Leonard, N. J., and Weber, G.,** Fluorescent modification of adenosine-containing coenzymes: biological activities and spectroscopic properties, *Biochemistry,* 19, 3499, 1972.
3. **Yoshioka, M. and Tamura, Z.,** Fluorimetric determination of adenine and adenosine and its nucleotides by high-performance liquid chromatography, *J. Chromatogr.*, 123, 220, 1976.
4. **Dünges, W.,** 4-Bromomethyl-7-methoxycoumarin as a new fluorescence label for fatty acids, *Anal. Chem.*, 49, 442, 1977.
5. **Lloyd, J. B. F.,** 4-Hydroxymethyl-7-methoxycoumarin: fluorescence characteristics and their variation in high-performance liquid chromatographic solvents, *J. Chromatogr.*, 178, 249, 1979.
6. **Lam, S. and Grushka, E.,** Labeling of fatty acids with 4-bromomethyl-7-methoxycoumarin via crown ether catalyst for fluorimetric detection in high-performance liquid chromatography, *J. Chromatogr.*, 158, 207, 1978.
7. **Dünges, W. and Seiler, N.,** High-performance liquid chromatographic separation of esters of 4-hydroxymethyl-7-methoxycoumarin, *J. Chromatogr.*, 145, 483, 1978.
8. **De Leenheer, A. P. and Gelijkens, C. F.,** Gas-liquid chromatography of pyrimidine and purine nucleosides as their N,O-permethyl derivatives, *Anal. Chem.*, 48, 2203, 1976.
9. **De Leenheer, A. P. and Cosyns-Duyck, M. C.,** Gas-liquid chromatography of some alkyl derivatives of 5-fluorouracil, *J. Chromatogr.*, 174, 325, 1979.
10. **Sun, S. R. and Chun, A. H. C.,** Determination of pentobarbital in serum by electron-capture GLC, *J. Pharm. Sci.*, 66, 477, 1977.

11. **Hulshoff, A. and Förch, A. D.,** Alkylation with alkylhalides as a derivatization method for the gas chromatographic determination of acidic pharmaceuticals, *J. Chromatogr.,* 220, 275, 1981.
12. **Pederson, C. J.,** New macrocyclic polyethers, *J. Am. Chem. Soc.,* 92, 391, 1973.
13. **Durst, H. D.,** Crown ether catalysis: "naked" anions as reactive intermediates in the synthesis of phenacyl esters, *Tetrahedron Lett.,* 2421, 1974.
14. **Durst, H. D., Milano, M., Kikta, E. J., Connelly, S. A., and Grushka, E.,** Phenacyl esters of fatty acids via crown ether catalysts for enhanced ultraviolet detection in liquid chromatography, *Anal. Chem.,* 47, 1797, 1975.
15. **Liotta, C. L., Harris, H. P., McDermott, M., Gonzalez, T., and Smith, K.,** Chemistry of "naked" anion. II. Reaction of the 18-crown-6 complex of potassium acetate with organic substances in aprotic organic solvents, *Tetrahedron Lett.,* p. 2417, 1974.
16. **Parker, A. J.,** Protic-dipolar aprotic solvent effects on rates of bimolecular reactions, *Chem. Rev.,* 69, 1, 1969.
17. **Barmes, S., Pritchard, D. G., Settine, R. L., and Geckle, M.,** Preparation and characterisation of permethylated derivatives of bile acids, and their application to gas chromatographic analysis, *J. Chromatogr.,* 183, 269, 1980.
18. **Jardine, I. and Weidner, M. M.,** Approach to the quantitative analysis of nucleotides by gas chromatograph-mass spectrometry, *J. Chromatogr.,* 182, 395, 1980.
19. **De Leenheer, A. P. and Gelijkens, G. F.,** Quantification of 5-fluorouridine in human urine by capillary gas-liquid chromatography with a nitrogen-selective detector, *J. Chromatogr. Sci.,* 16, 552, 1978.
20. **Parker, C. A. and Ree, S.,** Correction of fluorescence spectra and measurement of fluorescence quantum efficiency, *Analyst,* 85, 587, 1960.
21. **Twale, F. W. J. E. and Weber, G.,** Ultraviolet fluorescence of the aromatic amino acid, *Biochem. J.,* 65, 476, 1957.
22. **Wehry, W. L.,** *Fluorescence Theory: Instrumentation and Practice,* Guibault, G. G., Ed., Arnold, London, 1967, chap. 2.
23. **Majors, R. E.,** High performance liquid chromatography on small particle silica gel, *Anal. Chem.,* 44, 1722, 1972.
24. **Gehrke, C. W., Kuo, K. C., Davis, G. E., Suits, R. D., Waalkes, T. P., and Borek, E.,** Quantitative high-performance liquid chromatography of nucleosides in biological materials, *J. Chromatogr.,* 150, 455, 1978.
25. **Colonna, A., Russo, T., Esposito, F., Salvatore, F., and Cimino, F.,** Determination of pseudouridine and other nucleosides in human blood serum by high-performance liquid chromatography, *Anal. Biochem.,* 130, 19, 1983.
26. **Kuo, K. C., Gehrke, C. W., McCune, R. A., Waalkes, T. P., and Borek, E.,** Rapid quantitative high-performance liquid column chromatography of pseudouridine, *J. Chromatogr.,* 145, 383, 1978.
27. **Wu, A. T., Au, J. L., and Sadée, W.,** Hydroxylated metabolites of R,S-1-(tetrahydro-2-furanyl)-5-fluorouracil in rats and rabbits, *Cancer Res.,* 38, 210, 1978.
28. **Dünges, W., Bergheim-Irps, E., Straub, H., and Kaiser, R. E.,** Microtechniques for the gas chromatographic determination of barbiturates in small blood samples, *J. Chromatogr.,* 145, 265, 1978.

CONTINUOUS-FLOW UV-VIS SCANNING OF NUCLEIC ACID COMPONENTS SEPARATED BY HPLC

Leonard F. Liebes, Norberto Guzman, and Lon Alterman

INTRODUCTION

A common problem confronting investigators making use of the high sensitivity and resolution of HPLC is the identification of the separated compounds. While a variety of detectors is used for the measurement of biological components (UV/visible, fluorescence, electrochemical, radioactive), a good compromise that offers medium to high sensitivity combined with selectivity is the UV/visible absorbance detector, the most widely used HPLC detector in place today. When attempting to identify these compounds one is faced with the task of resolving complex mixtures of biological origin. Often the investigator cannot *a priori* know the identity of all individual components in a mixture he or she is separating. Standard methods of chemical analysis, when applied to the analysis of separated and collected components, such as nuclear magnetic resonance (NMR), infrared spectroscopy (IR), or elemental analysis, require large sample sizes (i.e., milligram amounts) that are far above separations commonly carried out by analytical HPLC. The use of mass spectroscopy (MS) alone or combined with liquid chromatography (LC), while highly sensitive for sample identification, requires a volatile sample as well as either pure solvents or solvents with volatile salts in LC-MS operation.

Discrimination of particular nucleic acid components is most readily accomplished on the basis of their spectral properties, although the use of radiolabeled compounds can also be employed. However, isotopes useful for nucleotide analysis such as ^{32}P pose a potential hazard to the investigator and have very short half-lives.

Comparison of retention times(s) with known standards is insufficient by itself as a means of identification. Absorbance ratios have been widely used as an aid in peak identification;[1-6] however, optimal utilization requires prior knowledge of the absorbance properties of the components in the mixture which is being separated. Spectral analysis provides for the chromatographer a sensitive means of identifying separated compounds. Information about the chemical nature of a compound can be derived from its absorption spectrum. Spectral data in first- and second-derivative functions may be analyzed with the help of microlaboratory computers along with software packages. This additional analysis is more sensitive to peak inflection points which ultimately leads to the discrimination of pure and impure components.

Currently, there are several means to achieve spectral identification of the components separated by HPLC. The technique of stopped-flow scanning allows examination of the entire spectrum of a chromatographic peak using a variable-wavelength HPLC detector.[5,7,8] While this method is highly useful when examining an unknown compound, a major drawback is the requirement to stop chromatographic flow during the wavelength scan. It is, in practice, difficult to achieve an instantaneous complete stoppage of flow due to inertia on pump pistons, incomplete seating of check valves, and plumbing leaks. In addition, it is often difficult to continue the analysis by resuming the flow, since the rest of the separation may be degraded.

CONTINUOUS-FLOW SCANNING

The selected option of peak scanning without interruption of solvent flow provides the chromatographer the ability to maintain peak quantification. These combined functions are

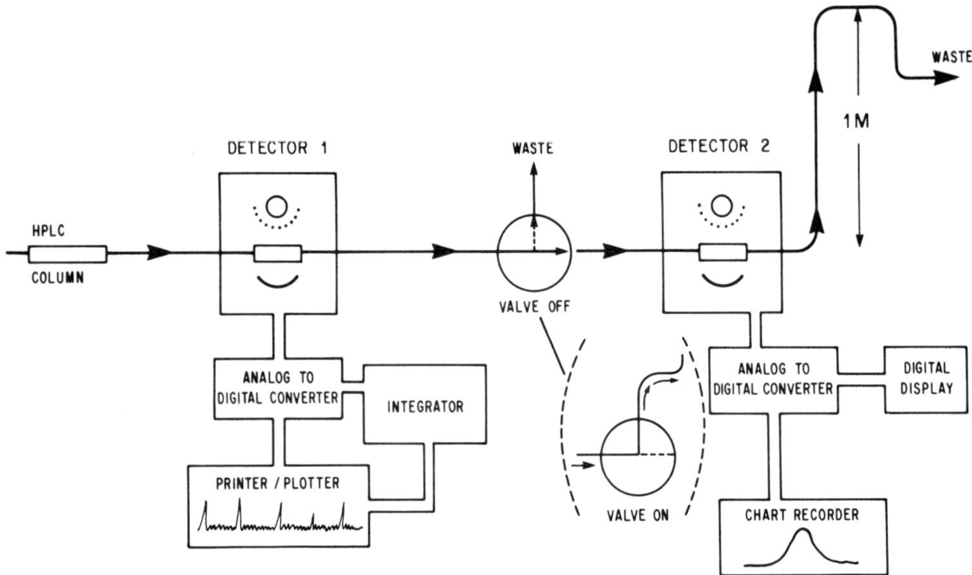

FIGURE 1. Post-column elution flow diagram for mechanical continuous-flow scanning using two detectors. The three-way valve between detector 1 and detector 2, in its inactivated state, allows the flow from the HPLC column to pass directly to detector 2, and out through a 1-m-high waste outlet. When current is applied to the valve, flow is diverted from detector 2 to waste leaving material trapped in the second detector. The 1-m height of the waste outlet of detector 2 coupled with the "on" configuration of the three-way valve prevents any diffusion of the trapped sample out of detector 2.

achieved through two different types of instrumentation, one of which functions through a combination of several instruments (mechanical continuous flow), while the other accomplishes the same features electronically (diode array detector, see following section). The mechanical version uses two detectors, one of which contains scanning capability, a three-way solvent control valve, a microprocessor-controller, a solenoid interface, data integrator, and chart recorder.

Figure 1 shows a flow diagram of the chromatographic system that is employed in the mechanical continuous-flow scanning. Since the outflow from the analytical column passes through detector 1 prior to detector 2 and then out to waste, quantification is continuously achieved by detector 1, while detector 2 is available for selected peak wavelength scanning. The use of a 1-m-high waste outlet connected to the control valve coupled with the "on configuration" of this valve serves to prevent any diffusion of the trapped sample out of detector 2 during the scanning sequence. A combination of two three-way valves operated in tandem would also achieve this "peak trapping". Given the requirement of high sensitivity operation, a background subtraction similar to that described for stopped-flow scanning may be required. Microprocessor control is required for an automated scanning sequence (Figure 2). For further details of the hardware involved in this technique, please refer to an earlier report.[9]

PHOTODIODE ARRAY HPLC DETECTOR

Diode array detectors (currently obtainable from Hewlett-Packard, LKB, Varian, and Shimadzu, etc.) acquire absorbance data as a function of wavelength and time. This is achieved through spectral dispersion of transmitted radiation of the sample onto an array of photodiodes. Photodiodes by nature have a fast response time. When they are coupled by an analog-to-digital converter and computer system, rapid processing of absorbance data

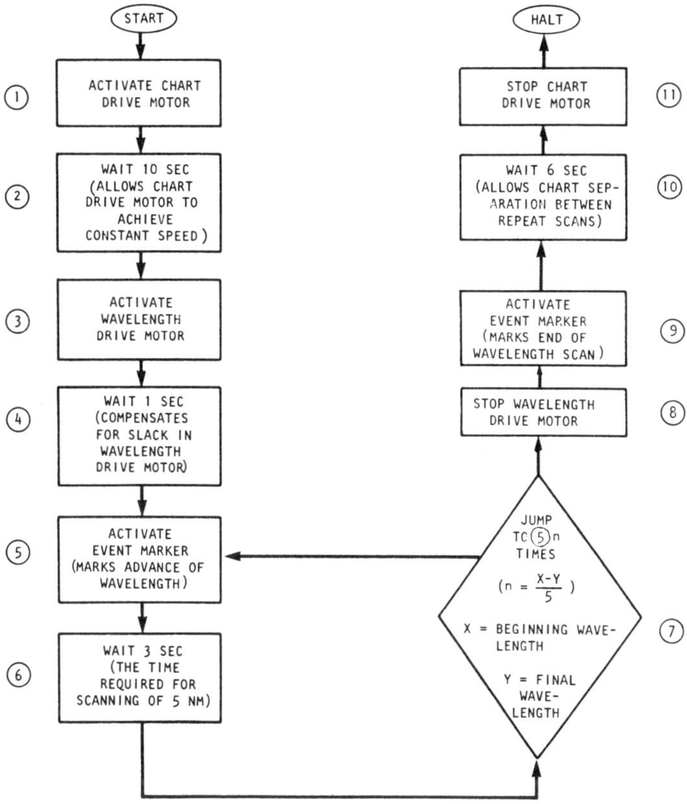

FIGURE 2. Flow chart of a typical program used to control an automatic scanning sequence with wavelength notation.

can be obtained. However, some of the available detectors lose some speed in the manner in which the data from the diode arrays are processed. The major advantage of the diode array detector is the ability to simultaneously monitor HPLC effluents at different wavelengths and to perform UV scans. Spectral resolution can be as good as 2 nm, although some units have resolution of 4 to 5 nm, and wavelengths can be monitored in the range from 190 to 600 nm. Data from the diode arrays are stored as separate addresses within a suitable computer memory bank or storage device. After a sample is injected into the column, a standard plot is obtained allowing quantification of the eluted peaks. While later, on an off-line basis, through the use of internal software processing packages, two- and, if desired, three-dimensional spectral scans can be obtained. Postanalysis data handling can allow the calculation of first- and second-derivative functions.

Figure 3 shows an elution profile obtained from an extract of human lymphocytes.[9] Selected peaks were sequentially scanned in the UV region during the course of the chromatographic separation. Figure 4 contains the separate spectral scans with a comparison of reference standards. The agreement between these two spectral scans is good with the exception of the far-UV portion of the spectra, where differential O_2 absorption and variations in the monochromators of the two instruments are more apparent. Data obtained with inosine monophosphate (IMP) (Figures 3 and 4B) show that closely resolved peaks can be subjected to continuous-flow UV scanning with satisfactory results.

Figure 5 shows the added flexibility obtainable with diode array scans. An entire chromatogram can be displayed as a function of selected wavelength components. This type of comparison can serve as a rapid screening of spectral differences amongst the various peaks prior to a more detailed analysis.

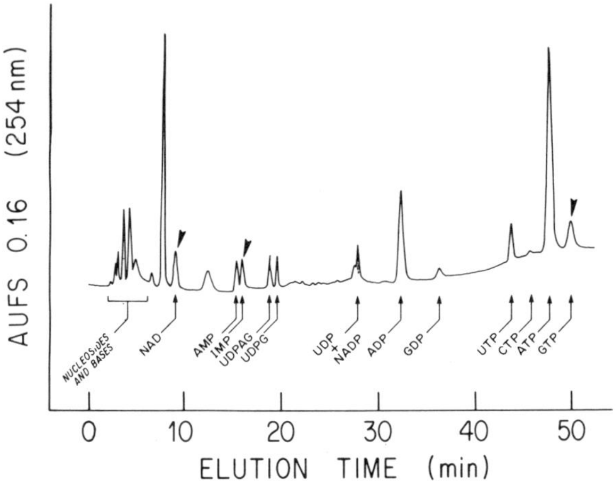

FIGURE 3. Separation of nucleotides on a strong anion exchanger column from 3 × 10^6 human lymphocytes. Chromatographic conditions are as described in a previous publication.[9] The marked peaks were subjected to peak trapping in the variable-wavelength detector during the same chromatographic separation using the mechanical continuous-flow scanning technique.

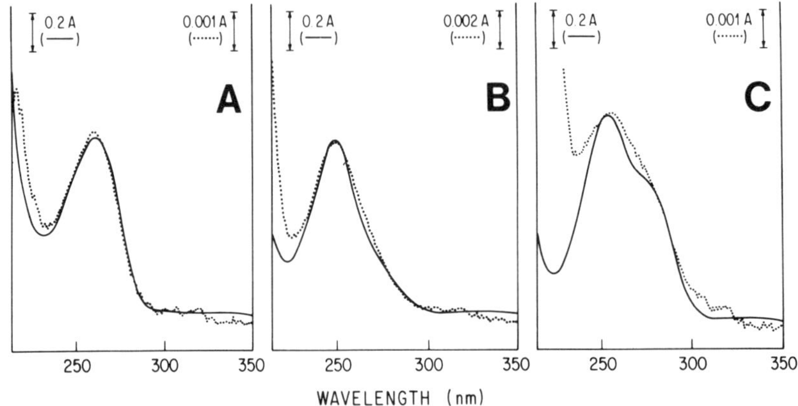

FIGURE 4. UV scans of trapped HPLC peaks by the mechanical continuous-flow scanning technique (. . .) and reference compounds (——) by a standard UV spectrometer. (A) NAD; (B) IMP; (C) GTP. Scanning range from 215 to 350 nm; absorbance range as indicated.

Figure 6 illustrates the ability of superimposing different diode array scans for comparative purposes. For example, scans of UDPAG and UDPG (the two comparable peaks which elute, respectively, at 13.4 and 14.4 min) show, as expected, the same spectral properties. The fidelity of diode array scans is dependent on the signal intensity of the eluting components. For example, when data are obtained on a major eluting component such as ATP (Figure 7), the spectral absorption is a smooth curve. However, when minor components are present, as with the UDPAG and UDPG peaks, the limited mass of these eluting samples yields less distinct spectral curves. The ability to rapidly store multiparameter data allows the end-user to process and/or reprocess data with considerable flexilibity, as long as there is sufficient data storage capability available. However, care must be taken in obtaining and subtracting the appropriate background scanning data.

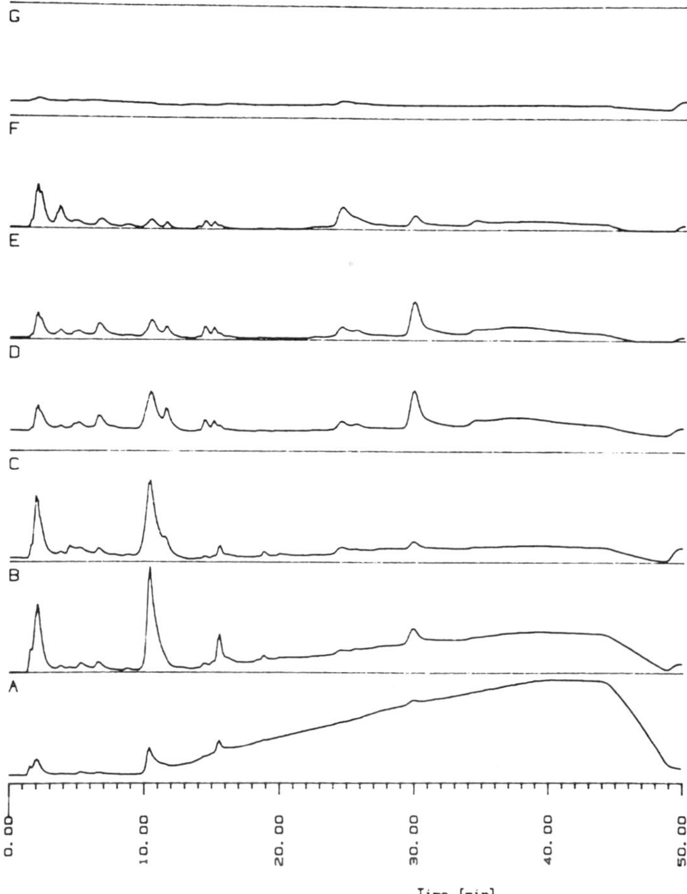

FIGURE 5. Multispectral chromatograms of separated nucleotide components from a lymphocyte nucleotide extract obtained from one analysis. The data are plotted after subtraction of spectral data obtained at 550 nm. The plots are, respectively, (A) 205 nm; (B) 214 nm; (C) 230 nm; (D) 254 nm; (E) 265 nm; (F) 280 nm; (G) 320 nm.

CONCLUSIONS

The simultaneous quantification and scanning of the same sample can be achieved by several methods. One technique employs mechanical continuous-flow scanning and makes use of two detectors. This technique obviates chromatographic artifacts which often accompany stopped-flow scanning and is useful to laboratories having access to several UV detectors, one of which has scanning capabilities.

Diode array detectors have the capability to simultaneously monitor at several wavelengths. The user can manipulate the data for maximum presentation and analysis of its spectral components. It does require a substantial investment both in terms of hardware and software components.[11] It should be noted that diode array detectors have wide bandwidths, most allow a 2-nm resolution, whereas monochromator-based instruments resolve to better than 1 nm. Because of the large amount of variables to be manipulated, a major investment of time is required for off-line reprocessing of data.

Spectral data obtained on a peak in the small volume employed by most commercial HPLC detectors (*circa* 10 to 30 $\mu\ell$) in a pathlength of 0.5 to 1 cm allow a sensitivity of

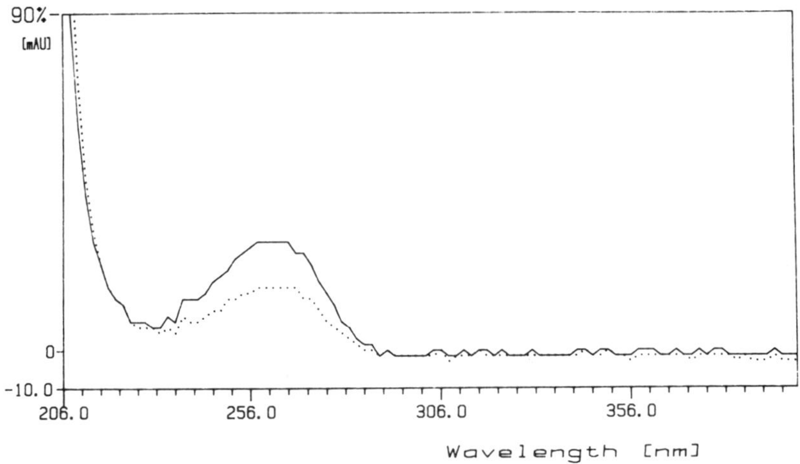

FIGURE 6. UV scans of UDPAG and UDPG obtained from diode array data from a lymphocyte nucleotide extract. The background reference was baseline data sampled prior to these two eluting peaks. (UDPAG, ———) (UDPG, . . .). Scanning range from 206 to 401 nm.

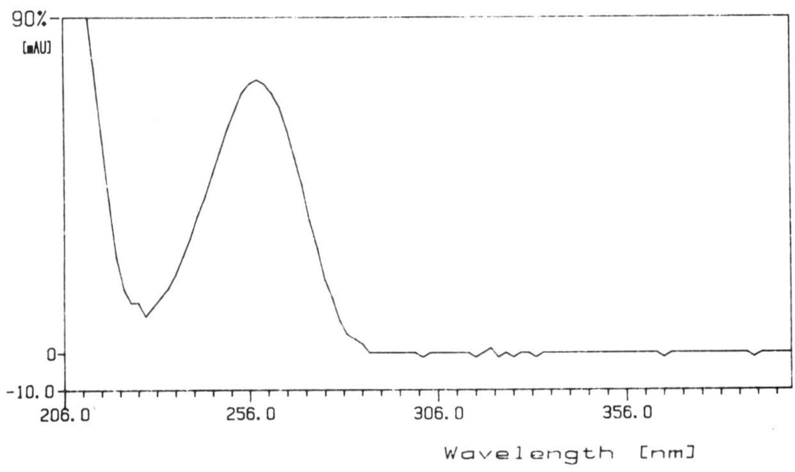

FIGURE 7. UV scan of ATP obtained from diode array data from a lymphocyte extract. The background reference and scanning range are the same as in Figure 6.

100 pmol for spectral characterization of typical nucleic acid components (e.g., with E_{mM} $\sim$ 10). Much more material is required to obtain a spectral scan in regular UV spectrophotometers using HPLC separated components. This could only be attempted with material that is collected from preparative scale HPLC separations. Using HPLC detectors, this apparent sensitivity can be increased to 10 pmol through the use of ultrafast HPLC 3-μm columns and low volume cells which cause minimal sample dilution.

Samples of biological origin are usually available in marginally adequate amounts which often lead to small yields of material for HPLC analysis. UV spectra provide a highly sensitive, generally nondestructive means for identification of nucleic acid components. When this is coupled with other information such as retention times or cochromatography with reference compounds,[12] these data can be used to corroborate the identity of nucleotides which may cochromatograph. Continuous-flow UV scanning thus is helpful in establishing the identity of nucleic acids, to access the purity of commercially obtained standards, and to distinguish among closely eluting chromatographic peaks. For maximum sensitivity and accurate interpretation of data, appropriate electronic or software-based background corrections are required for each spectral scan. When coupled with additional data processing such as first- and second-derivative calculations and/or three-dimensional plotting, the spectral purity of a separated component can also be ascertained with a high degree of certainty.

REFERENCES

1. **Krstulovic, A. M., Brown, P. R., and Rosie, D. M.,** Selective monitoring of polynuclear aromatic hydrocarbons by high pressure liquid chromatography with a variable wavelength detector, *Anal. Chem.*, 48, 1383, 1976.
2. **Krstulovic, A. M., Brown, P. R., and Rosie, D. M.,** Identification of nucleosides and bases in serum and plasma samples by reverese-phase high performance liquid chromatography, *Anal. Chem.*, 49, 2237, 1977.
3. **Krstulovic, A. M., Hartwick, R. A., Brown, P. R., and Lohse, K.,** Use of UV scanning techniques in the identification of serum constituents separated by high performance liquid chromatography, *J. Chromatogr.*, 158, 365, 1978.
4. **Yost, R., Stovenken, J., and Maclean, W.,** Positive peak identification in liquid chromatography using absorbance ratioing with a variable-wavelength spectrophotometric detector, *J. Chromatogr.*, 134, 73, 1977.
5. **Breter, H. J. and Zahn, R. K.,** The quantitative determination of metabolites of 6-mercaptopurine in biological material. II. Advantages of a variable-wavelength HPLC spectrophotometric detector for the determination of 6-Thio-purines, *J. Chromatogr.*, 137, 61, 1977.
6. **Baker, J. K., Skelton, R. E., and Ma, C. Y.,** Identification of drugs by high pressure liquid chromatography with dual wavelength ultraviolet detection, *J. Chromatogr.*, 168, 417, 1979.
7. **Krstulovic, A. M., Hartwick, R. A., and Brown, P. R.,** Reversed-phase liquid chromatographic separation of 3′5′-cyclic ribonucleotides, *Clin. Chem.*, 25, 235, 1979.
8. **Readman, J., Brown, L., and Rhead, M.,** Use of stop-flow ultraviolet scanning and variable-wavelength detection for enhanced peak identification and sensitivity in high-performance liquid chromatography, *Analyst (London)*, 106, 122, 1981.
9. **Liebes, L. F.,** Continuous-flow scanning of selected high-performance liquid chromatography peak components by microprocessor control. Application to analysis of extracts from human lymphocytes, *J. Chromatogr.*, 219, 255, 1981.
10. **Miller, J. C., George, S. A., and Willis, B. G.,** Multichannel detection in high performance liquid chromatography, *Science*, 218, 241, 1982.
11. **Klatt, L. N.,** Simultaneous multiwavelength detection system for liquid chromatography, *J. Chromatogr. Sci.*, 17, 225, 1979.
12. **Brown, P. R., Krstulovic, A. M., and Hartwick, R. A.,** Current state of the art in the HPLC analyses of free nucleotides, nucleosides and bases in biological fluids, *Adv. Chromatogr.*, 18, 101, 1980.

PLANAR CHROMATOGRAPHY AND ITS APPLICATIONS TO NUCLEIC ACID RESEARCH

M. F. Gonnord

INTRODUCTION

Bidimensional separations are based on the successive use of two different chromatographic developments along the two perpendicular axes of a paper sheet or a thin-layer plate. The first bidimensional separation was reported in 1944: the resolution of 15 of the 22 amino acids in the analyzed sample mixture was achieved in 5 days on a rectangular paper sheet.[1] Later, the combinations of one chromatographic and one electrophoretic separation[2] and of two electrophoretic displacements[3] were also described.

One can appreciate the importance of bidimensional techniques by considering that the components of a complex mixture can potentially be spread by the two developments from their point of application over a "two-dimensional volume" where more space is available for achieving their resolution than in a monodimensional system. As a first approximation, the resolution of a two-dimensional separation is the square of the resolution power on a column having the same length.

Up to now, bidimensional chromatography has been implemented only in thin-layer chromatograpy (TLC). Since the spots are spread over the entire plate, quantification is experimentally difficult and time consuming. For several decades, column elution chromatography has been the separation tool of choice for the analyst and bidimensional chromatography was less commonly used. With increasing need for separations of more complex mixtures, the separation power of monodimensional processes has reached a limit, and an increase in interest has appeared for bidimensional techniques[4] which are greatly benefitting from a better understanding of the different separation processes and of the explosive developments of detectors and data acquisition devices.

Nevertheless, bidimensional TLC has been constantly, if not always intensively, used during the past 40 years, and it has played an important role in the elucidation of several biochemical pathways.[5] In the last decade, a very powerful technique of separation of proteins has been developed by combining isoelectrofocusing and gel electrophoresis: over 1000 proteins were separated and protein maps were obtained which were characteristic of some pathological states.[6] Several groups have contributed to the development of "bidimensional gel electrophoresis".[6-8] More recently, bidimensional liquid chromatography was pioneered,[9] its perspectives outlined,[10] and the equipment developed and perfected.[11]

In this chapter the technical aspects of bidimensional separations will be presented and the potential of bidimensional methods will be discussed. The applications of bidimensional separation techniques to the analysis of nucleic acids and their derivatives will be reviewed.

BIDIMENSIONAL CHROMATOGRAPHY

In two-dimensional TLC $(TLC)^2$, a sample is spotted at the corner of a flat bed and the migration of the solvent is allowed in one direction, followed by another migration at a right angle to the first one. Successive feeding of the same solvent in both directions, when using a homogeneous sorbent layer, leads to a very slight increase in resolution corresponding to an increase of the distance of migration by a factor of $\sqrt{2}$. All spots align on a diagonal and no real profit is taken from the potentiality offered by the second dimension. In order to obtain more efficient systems, the entire chromatographic area must be used and appropriate different mobile phases must be selected for each successive development. Addition of a complexing agent to the second solvent system, change of pH, or solvent composition gives the technique a great flexibility.

An alternative is to use a chromatographic plate coated with two different adsorbents. A strip of phase I is coated along one edge of the plate adjacently to phase II with which the rest of the plate is coated. Such plates are commercially available. The plate is first developed along the strip by placing an edge perpendicular to it in a development chamber with an appropriate solvent. The unknowns are separated into a linear array of constituents. After drying and eventually reactivating phase II, the plate is placed into another developing chamber where a second solvent system migrates through the chromatographic bed, distributing the constituents of the analyzed mixture in a planar array of subconstituents. After drying the plate again, the subconstituents are to be detected.

The development in a given solvent of such a plate coated with two adjacent phases often alters the retention values obtained from the second development. Furthermore, since the wettability of both adjacent phases is different, the ascent of the first solvent through both supports is not homogeneous and distorted spots are obtained. In addition, if the second solvent is very different from the first one, its migration in the orthogonal direction may be irregular, resulting again in distorted spots.

This can be partially overcome by intermediate washing of the plate,[12] or by operating a first development on a separate strip coated with sorbent I and transferring the spot thus obtained onto sorbent II after clamping the strip on the main thin-layer plate and carrying out a second perpendicular development.[13] Another alternative is to coat a plate with a homogeneous slurry of two different sorbents. The composition of the eluents is then chosen in such a way that a different sorbent interacts with the solutes in each direction.[14] The composition of the so-called ''mixed phase'' thus obtained can be varied in order to optimize the bidimensional separation.

DETECTION IN TWO-DIMENSIONAL CHROMATOGRAPHY

The localization, characterization, identification, and quantification of spots are essential aspects of both TLC and $(TLC)^2$. Both TLC and $(TLC)^2$ employ identical techniques, the problem being more difficult with $(TLC)^2$ since the whole surface of the plate has to be evaluated. Quantification thus requires a long time and very sophisticated instrumentation which severely contrasts with the great simplicity of the separation method. In $(TLC)^2$, qualitative analysis is based on the determination of the exact position of the spot, and quantitative analysis relies on the integration of the spot profiles in both directions. Devices permitting the scanning and integration of the spot in one dimension by the use of a narrow slit cannot be used in $(TLC)^2$ where the spatial resolution of the scanner must be small compared with the spot size (about 10 to 20 times smaller than the spot base width). Practically, this means that, depending on the TLC conditions, it should not exceed 0.1 to 0.5 mm in diameter; accordingly, the number of parallel scans to be carried out is L/w, where L is the length of the development and w the slit width. L/w typically varies between 100 and 2000, which requires specific devices and software to align the scans, to detect, and to integrate the spots.

A primitive, simple, but tedious technique consists of scrapping the parts of the adsorbent layer containing each spot, recovering the corresponding compounds, and analyzing them. This method implies the preliminary localization of the spot and its further quantification by any conventional method. This method is slow and it is neither convenient nor quantitative.

Often, spots must be chemically revealed before detection by either spraying adequate chemicals or exposing the plate to gaseous reagents. A lack of uniformity in the chemical reaction will affect the detection performances. Fluorescent or colored spots can easily be detected under white or UV light. A nonvisible or nonfluorescent compound can be made detectable by complexation or addition of a radioactive marker, or by using the quenching effect of the compound on the fluorescence of an additive embedded in the chromatographic sorbent. The compound then appears as a dark spot on a brilliant background. Qualitatively, this technique allows an excellent determination of the spot position, but only a coarse measurement of its size and concentration. For quantitative analysis, fluorimetric or densitometric techniques are needed. Quantification is based on the decrease of light intensity transmitted or reflected by the spot compared to the surrounding background.[15] Only at very low concentration is the signal intensity proportional to the concentration of the detected species.[16] At higher concentrations, the light diffusion by the sorbent particles must be considered, and the relationship between the concentration and the signal is described by the Kubelka-Munk equation.[15] In addition, the response factor of a given compound depends on its coefficients of light adsorption and scattering. The monochromatic wavelength must be optimized for the compound of interest. Theoretically, transmission densitometry is more sensitive than reflection densitometry, although the former technique is more strongly affected by the nonhomogeneity of the layer.[17]

To overcome the problems arising from the noise level due to the lack of uniformity of the sorbent layer and poor light intensity, double beam instruments have been designed, which compensate for the background effects due to the layer. Reflection densitometry is more widely used, especially in the UV range.

In fluorimetry, a compound excited by an incident radiation emits a secondary radiation of a longer wavelength which is detected. Compounds which do not emit a secondary radiation and the nonspotted background do not fluoresce. The sensitivity of a measurement is optimized by adjusting the incident radiation. Commonly used wavelengths are 254 and 360 nm. When operated in the quenching mode, fluorescence consists of spiking the complete layer with a fluorescent phosphor additive. After development, the plate is irradiated with UV light and the sample spots appear as black spots. Fluorescence quenching is more sensitive in the transmission mode. Commercially available instruments can be operated in both densitometric or fluometric modes with the use of a set of appropriate mirrors and filters, usually with a monochromatic beam.

More sophisticated instruments permit the adjustment of both the excitation and emission wavelengths. A scanning optical beam illuminates the rectangle of the plate, the long dimension being perpendicular to the scanning direction. The signal is integrated over the whole slit length. Quantification is erroneous if the concentration is not homogeneous along the slit when the relationship between the light intensity and the spot concentration is not linear. However, the signal-to-noise ratio is improved by integration of the fluctuations caused by the sorbent heterogeneity. The integration of the signal over the slit permits to linearize the characteristic response of a single slit in terms of concentrations.[18]

Greater accuracy in quantification is obtained with dual wavelength densitometers; the fluctuations of the light source are balanced and a flying spot replaces the fixed slit aperture.

Video techniques employing a vidicon camera have permitted fast densitometric evaluation of thin-layer chromatograms. An attached signal-processing system allows to carry out interactive background subtraction or spot integration.[20,21]

Adapting monodimensional scanners to the detection of bidimensional thin-layer chromatograms is not an easy task, due to the lack of appropriate software to align parallel scans of the plate and to process them to obtain quantitative information. In addition, the time needed for scanning a complete plate is long.

Two-dimensional gel electrophoresis has catalyzed the development of two-dimensional detection devices which are also used for the TLC analysis of radioactive compounds. The principle is to obtain an autoradiograph which is then scanned with a very narrow visible light beam. Sophisticated data acquisition systems allow the filtering, data reduction, background subtraction, calibration, stretching, signal detection, and, also, identification by mapping.[22-24] Thousands of spots have been detected using this technique.

$(TLC)^2$ plates have been quantified by videodensitometric techniques. Fluorescence and adsorption measurements can be performed by using two-dimensional matrices of photosensitive pixels, with a sensitivity in the nanomole range.[21,25] The major advantage of the technique is its speed, and the main disadvantages are its cost and complexity.

A new form of planar chromatography has recently been conceived which combines two successive perpendicular displacements of the sample compounds in a layer by means of a solvent moving linearly under pressure in the sorbent bed.[9] The first elution is partially performed in order to leave the samples within the sorbent. Chromatography is then carried out perpendicularly to the first direction. Compounds are eluted from the square column and detected on-line using a photodiode array. The detector consists of a series of individual UV cells placed in parallel along the exit edge of the column and intercepting a monochromatic light passing through the detector cell constituted by a slot placed at the column exit. A time-dependent signal is obtained for each diode of the array. The resulting bidimensional chromatogram is recorded and processed through a data acquisition device and displayed on a color TV monitor. The column is operated under regular HPLC conditions and quantification is no longer affected by either the particle diffusivity or the plate characteristics. Another advantage of this technique is that it permits a large spot capacity with a reasonable time of analysis.[10]

PERFORMANCES OF BIDIMENSIONAL CHROMATOGRAPHY

TLC conditions have been greatly improved with the availability of fine particles of either plain or chemically bonded silica. Theoretical equations have been developed to relate the time of analysis, efficiency, and resolution to the parameters describing the plate, the chromatographic system, and separation, leading to a general scheme of optimization.[26-31] Although detailed calculations are beyond the scope of this chapter, it has to be recalled that careful choice of a TLC system is needed to perform a given separation: TLC can sometimes give better results than high performance TLC (HPTLC). As a quasi general rule, one must remember that small molecules having large diffusion coefficients are better analyzed on large particle plates, while large molecules with small diffusion coefficients are better separated on plates coated with small particles.[29]

The performance of a chromatographic column is classically defined in terms of efficiency. However, in TLC, the plate height concept is a complex function of the development length, the retention, and the characteristics of the solvent and the plate. The resolution power of a chromatographic system is best described in terms of peak capacity. In TLC, it is defined as the number of spots with a resolution of unity which can be placed between the origin and the solvent front. TLC permits good performances for easy separations.[32] A spot capacity between 10 and 20 is easily achieved. It can be increased by decreasing the solute diffusion coefficient and by increasing the solvent kinetic coefficient and the packing homogeneity of the plate. It is, however, nearly impossible to exceed a spot capacity of 30 in a monodirectional TLC. Also, the time of analysis increases dramatically with the desired peak capacity. A peak capacity of 15 can theoretically be achieved in less than 1 min on a 5-μm particle plate with a development length of 1.6 cm. In order to double the peak capacity, a 20-μm particle plate is needed and a development length of 30 cm with the time of analysis exceeding 1 hr. The same performance is obtained in about 4 min on a 5-cm-long HPLC column packed with 5-μm particles.[32] The unique advantage of TLC over liquid column chromatography is the possibility of carrying out two successive perpendicular developments using different retention mechanisms. The attained spot capacity is approximately the square of that achieved in unidirectional TLC. In bidimensional TLC, the spot capacity can be defined as the number of spots with a resolution of unity which can be placed in a rectangle delineated by the two perpendicular solvent fronts and the two parallels to these fronts passing through the center of the origin. For evaluation of the bidimensional spot capacity, the longitudinal and axial spreading of the spots in both developments must be accounted for. Guiochon et al.[33] have estimated the theoretical spot capacity in bidimensional TLC. They demonstrated that, when using a TLC plate coated with small particles, the bidimensional spot capacity ^2n decreases with decreasing particle size, because the development becomes very slow and the diffusion increases dramatically. When using large particles, an increase in the particle size and, thus, packing heterogeneity results in an increase of the flow velocity which broadens the spots and results in decreased bidimensional spot capacity. Again, the separation needs to be optimized by carefully selecting the plate and the solvent characteristics.[33] It nevertheless remains true that the spot capacity which can be attained using bidimensional TLC with a moderate analysis time is comparable and even larger than that which can be obtained with the best HPLC columns. Similar calculations have been carried out for the bidimensional column chromatographic technique, considering a development in the first dimension and elution in the second.[10] Peak capacities above 1000 are potentially obtainable with a 10 × 10-cm bidimensional column packed with 5-μm particles. With this technique, the overpressurization of the layer increases the packing homogeneity and permits a control of the solvent velocity which weakens the effect of diffusion on the peak capacity of the system.

Computer simulations have been used to optimize the bidimensional separations either by selection of an appropriate combination of solvents from monodimensional retention data,[34] or based on calculations of the retention behavior of the solutes with varying solvent composition.[35]

APPLICATIONS TO NUCLEIC ACID SEPARATIONS

The applications of planar chromatography to the separation of nucleic acids and their constituents have been mostly carried out by bidimensional TLC.

The first rapid bidimensional separation was realized by Randerath and Randerath.[36] There were 23 ribonucleotides separated on PEI cellulose layers. Di- and trinucleotides were resolved together with some nucleotide sugars. Step gradient elution with a LiCl solution was performed in the first dimension. After removal of LiCl, which would interfere with the subsequent elution, the chromatogram was developed in the perpendicular direction with formic acid-sodium formate buffers by a stepwise elution procedure. The separation was performed according to the charge of the solutes. The most negatively charged trinucleotides migrate the least on the PEI anion exchanger (Figure 1).

The separations of RNA hydrolysis products have been carried under similar bidimensional TLC conditions with successful resolution of the four major nucleotides found in RNA (guanine, adenine, uridine, and cytosine monophosphate and both 2′ and 3′ isomer of each of them). Detection was made under the UV light.[37] Barton et al.[38] have resolved nine pyridine- and adenine-containing nucleotides on PEI cellulose under similar conditions: a first development with 0.5 *M* sodium formate buffer, followed by a treatment with anhydrous methanol and a second perpendicular development in two zones with different LiCl concentrations.

PEI cellulose has been also used to resolve complex mixtures of nucleic acid bases, nucleosides, and nucleotides.[39] The separation of the bases and nucleotides from the nucleosides was realized by a first bidimensional analysis with a first development with water, and a perpendicular development using a butanol-methanol-water-ammonia (60:20:20:1) mixture. Nucleotides were resolved by step-gradient elution with LiCl in the first dimension, and formic acid-formate buffer in the second.

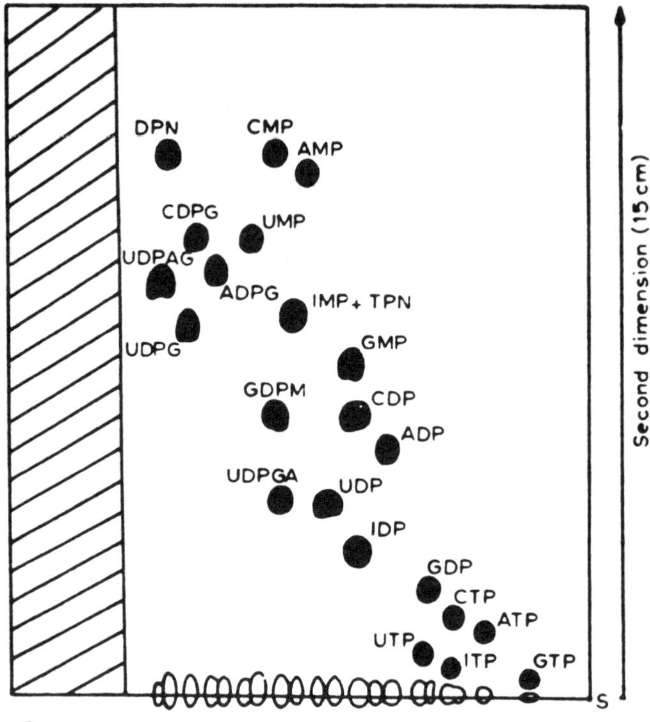

FIGURE 1. Two-dimensional anion exchange separation of nucleotides on PEI-cellulose. First dimension: successive developments in (a) 0.2 *M* LiCl, (b) 1.0 *M* LiCl, (c) 1.6 *M* LiCl. Second dimension: successive developments in (d) 0.5 *M* formate buffer (pH 3.4), (e) 2.0 *M* formate buffer, (f) 4.0 *M* formate buffer. CMP = cytosine 5′-monophosphate; AMP = adenosine 5′-monophosphate; CDPG = cytidine diphosphate glucose; UMP = uridine 5′-monophosphate; UDPAG = uridine diphosphate-*N*-acetyl-glucosamine; ADPG = adenosine diphosphate glucose; IMP = inosine 5′-monophosphate; GMP = guanosine-5′-monophosphate; GDPM = guanosine diphosphate mannose; CDP = cytosine 5′-disphosphate; ADP = adenosine 5′-diphosphate; UDPGA = uridine diphosphate glucuronic acid; UDP = uridine 5′-diphosphate; IDP = inosine 5′-diphosphate; GDP = guanosine 5′-diphosphate; CTP = cytosine 5′-triphosphate; ATP = adenosine 5′-triphosphate; UTP = uridine 5′-triphosphate; ITP = inosine 5′-triphosphate; GTP = guanosine 5′-triphosphate; DPN = nicotinamide adenine dinucleotide; TPN = nicotinamide adenine dinucleotide phosphate. (Reproduced from Randerath, E. and Randerath, K., *J. Chromatogr.*, 16, 126, 1964. With permission.)

Another procedure[40] involves two preliminary chromatographies on silica gel plates for subsequent resolution of arabinosyl and ribosyl nucleosides. Purine nucleotides were then separated by bidimensional TLC on PEI cellulose sheets with a LiCl step gradient elution in the first direction followed by a perpendicular elution with an ethanol-2 *M* LiCl (1:1) mixture saturated with H_3BO_3 and neutralized.[40] All the components were resolved in a single chromatogram, with the exception of the nucleoside monophosphates of hypoxanthine and guanine, which were still overlapping. Cellular nucleotides were separated on PEI cellulose plates by combining a separation by charge in the first direction and by base in the second one.[40a] A bidimensional separation of cellular nucleotides with 0.9 *M* guanidine HCl (pH 6.5) for the first development and 74 g $(NH_4)_2SO_4$, 0.4 g NH_4HSO_4, 4 g of disodium

EDTA, 100 mℓ H_2O (pH 3.5) in the second direction was also detailed.[41] (Figure 2). Autoradiography was used for detection. An optimized $(TLC)^2$ separation of quinone-induced spots is also described.

Cellulose was often selected as a sorbent for various separations. Bieleski[42] separated phosphate ester extracts by $(TLC)^2$ with two successive elutions in each direction with a basic solvent system for the first direction and an acidic solvent system in the second dimension. The phosphate ester pattern obtained on the TLC plate is very similar to the one obtained with the same solvent system by two-dimensional paper chromatography. Another planar separation was obtained by this author by combining one TLC development with electrophoresis (Figure 3).

Separations of other phosphate esters were described by Cole and Ross.[43] Cellulose was also used for the bidimensional separation of 2′- and 3′-monophosphates,[44] 2′-*O*-methyl-nucleotides,[45] and purine and pyrimidine bases.[13] Also, purines, pyrimidines, and their nucleosides in urine have been screened by this technique.[46] Cases of specific nucleoside deficiencies were discussed. The pH of the eluent affects the degree of solute dissociation and thus plays an important role in this separation. In most cases, the bidimensional development is achieved by using an acidic solvent in one direction and a basic solvent in the other one. It is also possible to couple the use of a basic solvent with the use of a salt to specifically affect the nucleotide retention.

Intercellular purine interconversion products were analyzed on a Polygram® CEL 400 thin-layer sheet by three successive developments with methanol, then methanol-propanol 25% NH_3-1% EDTA (3:9:2:6) twice in the first dimension, and a unique perpendicular development with isobutyric acid-25% NH_3-1% EDTA (200:9:114). Autoradiographic detection permitted the detection of extremely low amounts of purine derivatives.[47] Pataki resolved a mixture of 44 bases, nucleosides, deoxynucleosides, 2′-, 3′-, and 5′-nucleoside monophosphates, nucleoside di- and triphosphates, and nucleotide sugars on cellulose with an *n*-propanol-25% ammonia-water (5:3:1) mixture in the first direction, and an isopropanol-saturated ammonium sulfate-water (2:79:19) mixture in the second.[48] Triphosphates exhibited strong retention in the first system, while the ionic strength effect modified their retention in the second eluent system without affecting the position of nucleosides and bases in the chromatogram (Figure 4).

Munns et al.[49,50] separated 22 methylated tRNA purine and pyrimidine bases using a mixed phase plate (60% microcrystalline cellulose-40% silica gel) with ethyl acetate-methanol-water-88% formic acid (100:25:20:1) mixture in the first dimension, and acetonitrile-ethyl acetate-2 propanol-1 butanol-58% ammonia-water (40:30:20:10:22:5) in the second. The relative percentage of both phases was found to be of critical importance.[49] Solutes with primary amine groups like adenine, guanine, and cytosine exhibited lower retention in the formic acid than in the ammonia system. These differences resulted from the interaction with silica, the adsorption of the primary amines being also decreased by their methylation. At basic pH values, adsorbent-solute interactions were weaker, for both the bases and the silica phase were negatively charged.[50] Using a pure silica gel HPTLC plate, Das separated adenine and guanine nucleotides and nucleosides in the picomolar concentration range by varying the solvent pH.[51]

Separations of oligonucleotides and nucleic acids by bidimensional chromatography can be achieved in different ways. Basically, when two chromatographic developments are involved, PEI-cellulose or cellulose are chosen as a stationary phase, more seldomly silica gel. The main novelty is the possibility of combining two fundamentally different techniques such as chromatography and gel-electrophoresis.

Bidimensional TLC was used for the separation of complete RNase T_1 digestion products of 5s RNA on PEI cellulose plates[52] (Figure 5). In the first dimension, a two-step development in pH 3.4 solvent was realized with successive use of 1.4 and 1.8 *M* solutions of lithium

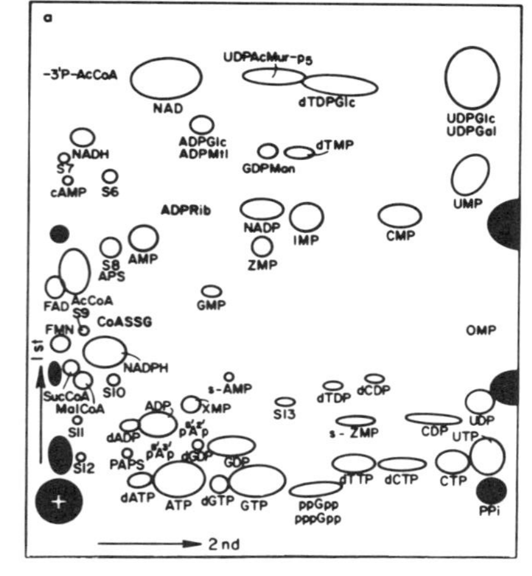

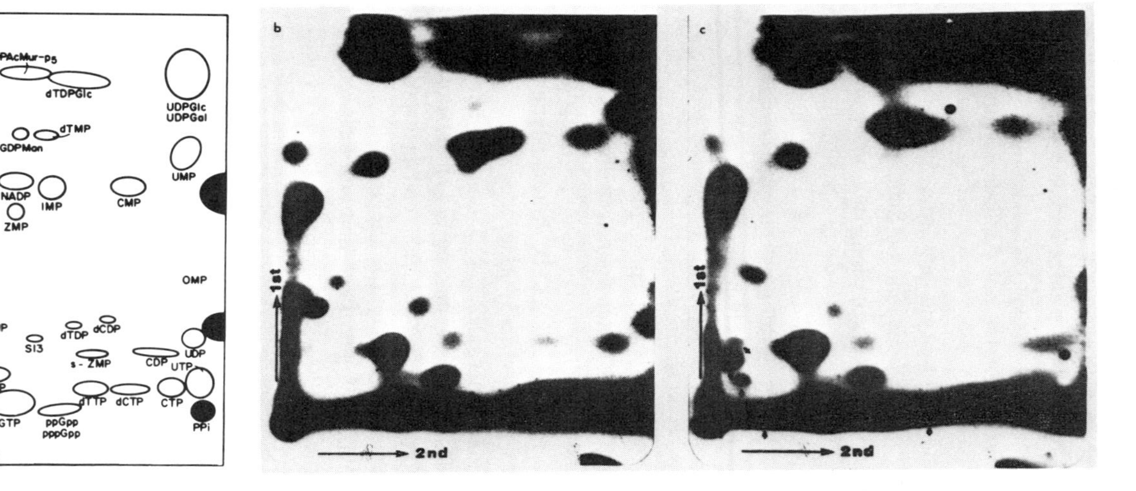

FIGURE 2. (TLC)[2] separation of cellular nucleotides. 0.9 *M* guanidine HCl (pH 6.5) was used as the first dimension as solvent, and 74 g of $(NH_4)_2SO_4$, 0.4 g of $(NH_4)HSO_4$, 4 g of disodium EDTA, and 100 mℓ of H_2O (pH 3.5) were used for the second-dimension solvent. (a) Identities of metabolites resolved with this two-dimensional system, including 3′-phosphoadenosine 5′-phosphosulfate (PAPS), acetyl coenzyme A (AcCoA), succinyl coenzyme A (SucCoA), malonyl coenzyme A (MalCoA), and uridine 5′-diphospho-*N*-acetylmuramyl pentapeptide (UDP-AcMur-p_5). Also shown are the chromatographic locations of ADP-Rib, adenosine 5′-phosphosulfate (APS), coenzyme A-gluthathione disulfide (CoASSG), 3′-dephospho coenzyme A(-3′ P-CoA), orotidine 5′-monophosphate (OMP), $p^{5'}A^{2'}p$. S6 to 13 represent phosphorylated compounds present in normal cells that have not been identified. The darkened spots represent compounds that do not adsorb to charcoal. Autoradiograms exposed from this two-dimensional separation of a ^{32}P-labeled extract of *Salmonella typhimurium* before (b) and 30 min after (c) addition of ACDQ (5 μg/mℓ) are shown. The arrows indicate the appearance of induced phosphorylated metabolites. (Reproduced from Lee, P. C., Bochner, B. R., and Ames, B. N., *J. Biol. Chem.*, 258, 6827, 1983. With permission.)

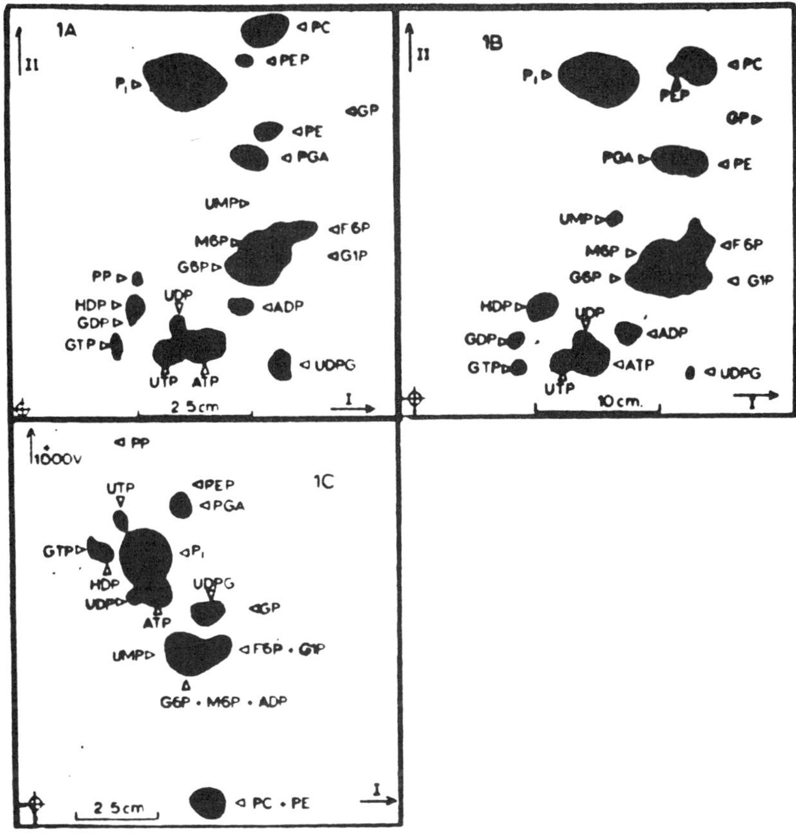

FIGURE 3. Separation of phosphate esters. (1A) By $(TLC)^2$ in solvent I (twice) and solvent II (twice). Sample, 1 μℓ phosphate ester extract (equivalent to 4 mg *Brassica pekinensis* vascular bundles) with activity 400 dps (3000 cpm). Autoradiographed 30 hr; (1B) by two-dimensional paper chromatography in solvent I and solvent II. Sample, 10 μℓ extract, activity 4000 dps. Autoradiographed 72 hr; (1C) by TLC in solvent I (twice) and by electrophoresis at pH 3.6 (1000 volts, 35 ma, 16 min). P_i, inorganic phosphate; PC, phosphatidyl choline; PEP, phosphoenolpyruvate; PGA, 2- + 3-phosphoglyceric acid; PE, phosphatidyl ethanolamine; F6P, fructose 6-phosphate; M6P, mannose 6-phosphate; G6P, glucose 6-phosphate; AMP, adenosine 5′-monophosphate; ADP, adenosine diphosphate; ATP, adenosine triphosphate; UMP, UDP, and UTP, uridine 5′-mono-, di-, and triphosphate; UDPG, uridine diphosphoglucose; GDP and GTP, guanosine di- and triphosphate; HDP, hexose diphosphate; PP, pyrophosphate. Solvent I: *n*-propanol/ammonia/water 6:3:1 v/v + 2g EDTA per liter; solvent II: *n*-propylacetate/90% formic acid/water 11:5:3 v/v. (Reproduced from Bieleski, R. L., *Anal. Biochem.*, 12, 230, 1965. With permission.)

formate. The lower concentration solvent gave a better separation of smaller oligoribonucleotides, and the higher concentration solvent a better separation of the larger ones. The migration of solutes depended on their charge and nucleotide composition. In the second dimension, the separation obtained with a 0.8 *M* lithium chloride solution in Tris base solvent (pH 8.0) was a function of solute molecular weights. The same technique was used for more complex mixtures of partial RNase T_1 fragments of 5s RNA with continuous gradient elution. The entire apparatus was maintained at 60°C in order to allow for the dissociation and separation of polynucleotide complexes into individual components and also to obtain a better resolution by sharpening the spots. Sequential analysis of oligonucleotide fragments of tRNA employing tritium postlabeling was described by Chen and Roe.[53] The interpretation of observed mobility shifts permitted the determination of the complete nucleotide sequence

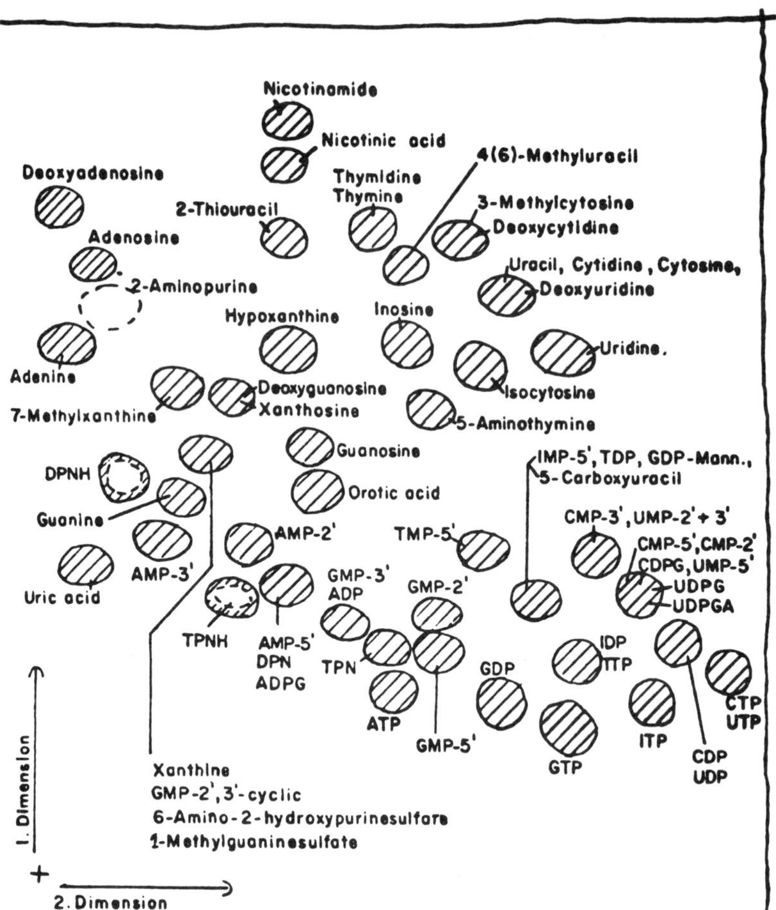

FIGURE 4. Two-dimensional separation of nucleo-derivatives on purified cellulose layers. First dimension, *n*-propanol-25% ammonia-water (6:3:1); second dimension, isopropanol-saturated ammonium sulfate-water (2:79:19). Absorption spots are hatched, fluorescence spots are surrounded by broken lines. (Reproduced from Pataki, G., *J. Chromatogr.*, 29, 126, 1967. With permission.)

of large oligonucleotide fragments. The procedure was applied to the complete nucleotide sequence analysis of several RNase T_1 and RNase A digestion products from human placenta tRNA.

Two-dimensional fractionation of the 5′-triphosphate terminal ends of RNAs made from various DNA templates was described by Miller and Burgess:[54] separation was carried out on PEI cellulose by using a 2 *M* LiCl-0.01 *M* EDTA (pH 6.5) solution in the first dimension and a 1.5 *M* LiCl-1.8 *M* formic acid-0.005 *M* EDTA (pH 2.0) solution in the second. The charges of both the phosphate and base molecules affected their migration.

The separation of about ten blocked methylated mRNA 5′-termini was achieved on a cellulose plate by combining different ionic equilibria in each direction and their effect on solute migration patterns.[55] Two different bidimensional separations were described. They allowed rapid qualitative and quantitative analysis of complex mixtures of mRNA cap structures on the basis of their methyl group content and base composition.

Silica gel sheets were used to achieve the bidimensional separation of apurinic DNA oligonucleotides with an isopropanol-water (75:25) solvent mixture in the first dimension and a butanol-acetic acid-0.5% ammonia (90:30:40) mixture in the second. Thymus DNA

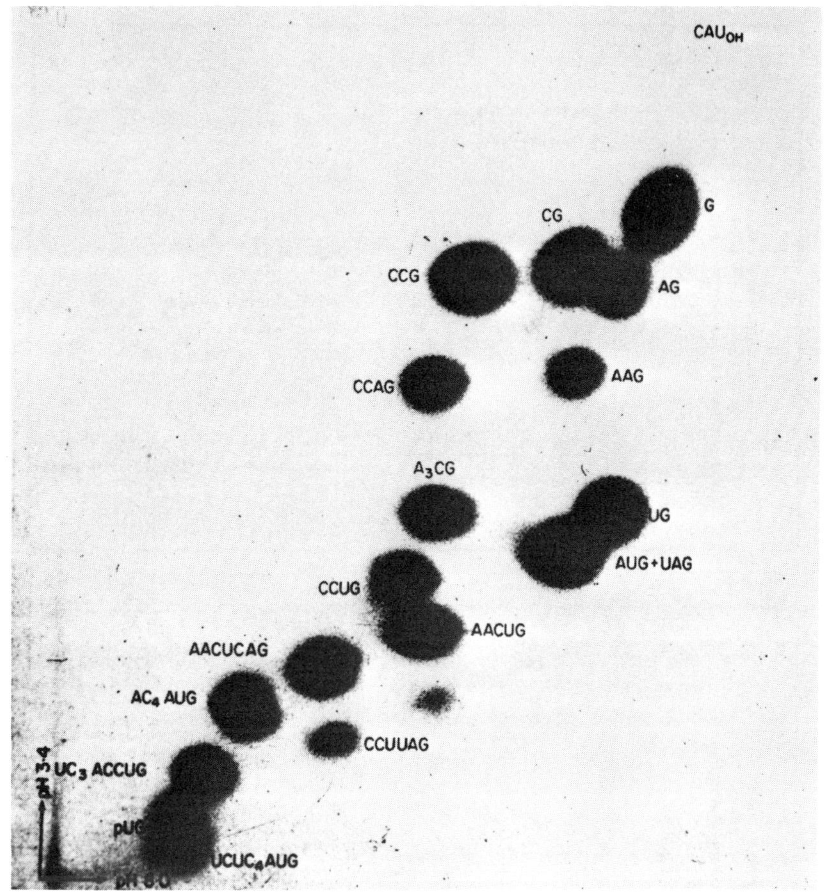

FIGURE 5. A two-dimensional separation of the complete RNase T_1 digestion products of 5s RNA on PEI-cellulose plates (20 × 20 cm). The first-dimensional separation was carried out at room temperature using 1.4 and 1.8 *M* lithium formate (pH 3.4) in 7 *M* urea; the second, also at room temperature, used the lithium chloride (pH 8.0, 0.8 *M*) system. The 3′-end product, CAU_{OH}, was the only product not defined as a sharp spot by this technique. Identities of the products were confirmed by elution and hydrolysis to mononucleotides. (Reproduced from Mirzabekov, A. D. and Griffin, B. E., *J. Mol. Biol.*, 72, 633, 1972. With permission.)

was separated into 21 spots, among which 16 were identified as nucleotides by UV spectrophotometry after enzymatic treatment.[56]

The bidimensional separation can be hastened by coupling electrophoresis with TLC. In this scheme, Bergquist mapped the enzymatic digests of ribonucleic acids by separating 17 oligo- and mononucleotides on a 20 × 20-cm cellulose layer.[57] Electrophoresis was primarily run at pH 2.5 at 50 V/cm. Two successive chromatographic developments were then applied at a right angle to the electrophoretic dimension by consecutive uses of *tert*-butanol-0.08 *M* formic and isoamyl alcohol (50:5:2 v/v) followed by *n*-butanol-water (86:14) (Figure 6). The last development did not alter the positions of the nucleotides, but removed the excess of fluorescent material contaminating the mono- and dinucleotides. Spots were located under the UV light and identified on the basis of UV spectra and chromatographic retentions compared with those of known oligonucleotides. A similar procedure has been used for mapping oligonucleotides from digests of RNA with pancreatic RNase, and it was also applied to hydrolysates obtained with RNase T_1 or U_2.[58] Reproducible and characteristic fingerprints were obtained from which the structure of an oligonucleotide was deduced from its position on the map.

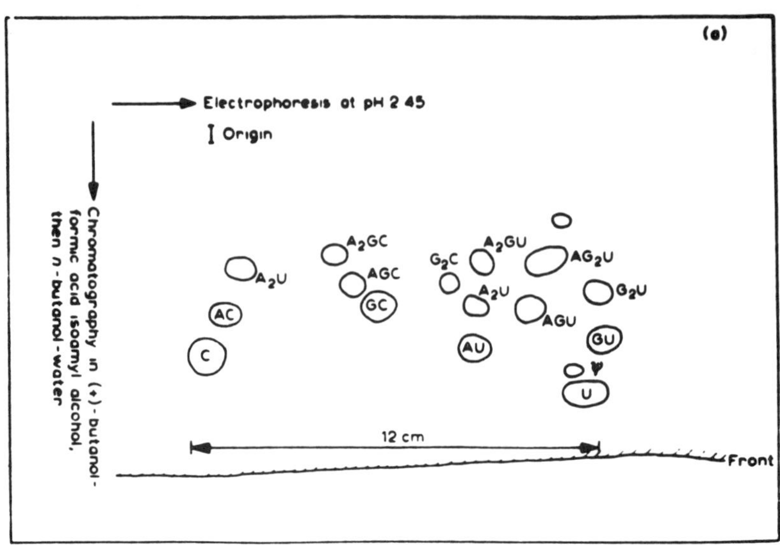

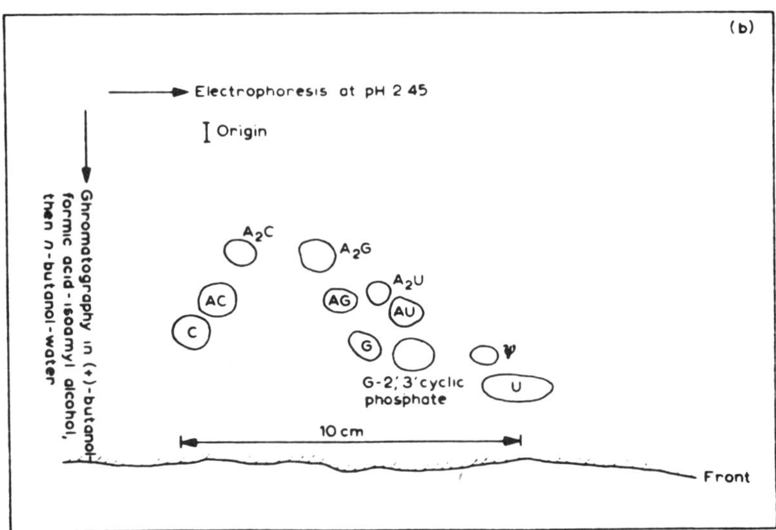

FIGURE 6. Separation of enzymatic hydrolysis products of sRNA by thin-layer electrophoresis and thin-layer chromatography. (a) Separation of mono- and oligonucleotides from a pancreatic ribonuclease digest of *E. coli* sRNA; (b) separation of mono- and oligonucleotides resulting from the action of pancreatic ribonuclease plus T_1 ribonuclease on *E. coli* sRNA. (Reproduced from Bergquist, P. L., *J. Chromatogr.*, 19, 615, 1965. With permission.)

In addition, the combination of gel electrophoresis and TLC has been applied to RNA sequencing studies.[45,59]

Fractionation by gel electrophoresis separates 3′-terminal RNA fragments. It is then sufficient to determine the 5′-terminal base of each fragment to obtain the RNA sequence. The band is thus excised from gel, transferred in thin layer of DEAE cellulose by blotting, and *in situ* hydrolyzed with takadiastase RNAse. A thin-layer development is then performed with 15 m*M* sodium citrate buffer (pH 3.1). The enzyme hydrolyzes one nucleotide at a time and the resulting pNp spots correspond to the nucleotide sequence in the RNA molecule. The resulting pattern is detected by autoradiography with direct reading of the RNA sequence.[59] The major 5′-terminal RNase T_1 oligonucleotides derived from human adenovirus type 2 early regions E_{1a} and E_{1b} were characterized and sequenced by resolving the oligonucleotides by a two-dimensional paper electrophoresis-homochromatography technique, followed by the characterization of the 5′-terminal methylated oligonucleotides by nuclease P_1 digestion and appropriate bidimensional TLC and RNase T_2 digestion and another $(TLC)^2$. Sequencing was realized by partial nuclease P_1 digestion and electrophoresis homochromatography.

CONCLUSION

As exemplified by its applications to nucleic acid research, planar chromatography is a rather simple tool for identifying nucleic acid constituents and for determining RNA sequences.

A better understanding of the TLC process, together with the availability of high performance plates, and new developments in the coupling of separation techniques should favor a regain of interest in this technique.

The main disadvantage of bidimensional systems is the difficulty of quantifying a large number of spots. However, this can be partially overcome by using new detector technology.

Nevertheless, its great efficiency and the ability to resolve very different compounds using two different retention mechanisms make bidimensional chromatography a very powerful and rapid separation technique.

One can expect that instrumentation for bidimensional separation techniques will be the subject of new developments in the near future.

REFERENCES

1. **Consden, R., Gordon, A. H., and Martin, A. J. P.,** Qualitative analysis of proteins: a partition chromatographic method using paper, *Biochem. J.*, 38, 244, 1944.
2. **Haugaard, G. and Kroner, T. D.,** Partition chromatography of aminoacids with applied voltage, *J. Am. Chem. Soc.*, 70, 2135, 1948.
3. **Durrum, E. L.,** Continuous electrophoresis and ionophoresis on filter paper, *J. Am. Chem. Soc.*, 73, 4875, 1951.
4. **Giddings, J. C.,** Two-dimensional separations: concept and promise, *Anal. Chem.*, 56, 1258A, 1984.
5. **Pocchiari, F. and Rossi, C.,** Quantitative radio paper chromatography, *Chromatogr. Rev.*, 4, 1, 1961.
6. **O'Farrell, P. H. J.,** High resolution two-dimensional electrophoresis of proteins, *J. Biol. Chem.*, 250, 4007, 1975.
7. **Anderson, N. G. and Anderson, N. L.,** Analytical techniques for cell fractions. XXI. Two-dimensional analysis of serum and tissue proteins: multiple isoelectric focusing, *Anal. Biochem.*, 85, 331, 1978.
8. **Jellum, E.,** Multicomponent analyses of human body fluids and tissues in health and diseases using capillary chromatography-mass spectrometry and high resolution two-dimensional electrophoresis, in *Advances in Chromatography*, Zlatkis, A., Ed., Elsevier, Amsterdam, 1982, 139.
9. **Guiochon, G., Gonnord, M. F., Zakaria, M., Beaver, L. A., and Siouffi, A. M.,** Chromatography with a two-dimensional column, *Chromatographia*, 17, 121, 1983.
10. **Guiochon, G., Beaver, L. A., Gonnord, M. F., Siouffi, A. M., and Zakaria, M.,** Theoretical investigation of the potentialities of the use of a multidimensional column in chromatography, *J. Chromatogr.*, 255, 415, 1983.
11. **Gonnord, M. F. and Guiochon, G.,** On-line U.V. detection for two-dimensional elution overpressured chromatography, in *TLC with Special Emphasis on Overpressured Liquid Chromatography (OPLC)*, Tyihák, E., Ed., Labor Mim Publ., Budapest, Hungary, 1986, 241.
12. **Randerath, K. and Randerath, E.,** Resolution of complex nucleotide mixtures by two-dimensional anion-exchange thin-layer chromatography, *J. Chromatogr.*, 16, 111, 1964.
13. **Bond, P. S.,** A two-dimensional thin-layer chromatographic technique for the separation of admixtures of 5-bromouracil and DNA bases, *J. Chromatogr.*, 34, 554, 1968.

14. **Righezza, M., Siouffi, A. M., and Guiochon, G.,** Phases stationnaires mixtes pour la chromatographie en couche mince bidimensionnelle, *Analusis,* 10, 535, 1984.
15. **Frei, R. W.,** Reflectance spectroscopy in thin-layer chromatography, *Prog. Thin-Layer Chromatogr. Relat. Methods,* 2, 1, 1971.
16. **Pollak, V.,** Linear relationship between concentration and optical response of thin-layer chromatograms, *J. Chromatogr.,* 152, 201, 1978.
17. **Goldman, J. and Goodall, R. R.,** Quantitative analysis on thin-layer chromatograms: a theory for light adsorption methods with an experimental vertification, *J. Chromatogr.,* 32, 24, 1968.
18. **Touchstone, J. C., Levin, S. S., and Murawec, T.,** Spectrodensitometry of thin layer chromatograms, in *Quantitative Thin-Layer Chromatography,* Touchstone, J. C., Ed., Wiley Interscience, New York, 1979, 1.
19. **Pollak, V.,** Computers in quantitative thin-layer chromatography, *J. Liq. Chromatogr.,* 3, 1881, 1980.
20. **Kerenyi, G., Pataki, T., and Devenyi, T.,** Hungarian Patent 170287, 1976.
21. **Devenyi, T. and Pungor, S.,** High speed videodensitometry: a general overview, *Hung. Sci. Instrum.,* 50, 3, 1980.
22. **Anderson, N. L., Taylor, J., Scandora, A. E., Coulter, B. P., and Anderson, N. G.,** The TYCHO system for computerized analysis of two-dimensional gel protein mapping data, *Clin. Chem.,* 27, 1807, 1981.
23. **Lester, E. P., Lemkin, P. F., and Lipkin, L. E.,** New dimensions in protein analysis, *Anal. Chem.,* 53, 390A, 1981.
24. **Thorsrud, A. K., Haugen, H. F., and Jellum, E.,** High resolution two-dimensional electrophoresis (iso-DALT) in clinical chemistry: comparison with routine electrophoretic and immunological methods, in *Electrophoresis 79: Methods and Theories, Biochemical and Clinical Applications,* Vol. 53, Radola, B. J., Ed., Walter de Gruyter, Berlin, 1981, 1357.
25. **Gianelli, M. L., Callis, J. B., Anderson, N. H., and Christian, G. D.,** Thin layer chromatography with *in situ* multichannel image detection of fluorescent compounds, *Anal. Chem.,* 53, 1357, 1981.
26. **Guiochon, G., Siouffi, A. M., Engelhardt, H., and Halász, I.,** Study of the performances of thin-layer chromatography. I. A phenomenological approach, *J. Chromatogr. Sci.,* 16, 152, 1978.
27. **Guiochon, G. and Siouffi, A. M.,** Study of the performances of thin-layer chromatography. II. Band broadening and plate height equations, *J. Chromatogr. Sci.,* 16, 470, 1978.
28. **Guiochon, G. and Siouffi, A. M.,** Study of the performances of thin-layer chromatography. III. Flow velocity of the mobile phase, *J. Chromatogr. Sci.,* 15, 598, 1978.
29. **Siouffi, A. M., Bressolle, F., and Guiochon, G.,** Optimization in thin-layer chromatography. Some practical considerations, *J. Chromatogr.,* 209, 129, 1981.
30. **Guiochon, G., Siouffi, A. M., and Bressolle, F.,** Study of the performances of thin-layer chromatography. IV. Optimization of experimental conditions, *J. Chromatogr. Sci.,* 17, 368, 1979.
31. **Snyder, L. R.,** *Principles of Adsorption Chromatography,* Marcel Dekker, New York, 1968, 99.
32. **Guiochon, G. and Siouffi, A. M.,** Study of the performances of thin-layer chromatography. Spot capacity in thin-layer chromatography, *J. Chromatogr.,* 245, 1, 1982.
33. **Guiochon, G., Gonnord, M. F., Siouffi, A. M., and Zakaria, M.,** Study of the performances of thin-layer chromatography. Spot capacity in two-dimensional thin-layer chromatography, *J. Chromatogr.,* 250, 1, 1982.
34. **Gonnord, M. F., Levi, F., and Guiochon, G.,** Computer assistance in the selection of the optimum combination of systems for two-dimensional chromatography, *J. Chromatogr.,* 264, 1, 1983.
35. **Johnson, E. K. and Nurok, D.,** Computer simulations as an aid to optimizing continuous development two-dimensional thin layer chromatography, *J. Chromatogr.,* 302, 135, 1984.
36. **Randerath, E. and Randerath, K.,** Ion-exchange chromatography of nucleotides on poly (ethyleneimine)-cellulose thin layers, *J. Chromatogr.,* 16, 126, 1964.
37. **Santini, G. and Ulrich, V.,** Resolution of RNA hydrolysis products by two-dimensional ion-exchange thin-layer chromatography, *J. Chromatogr.,* 49, 560, 1970.
38. **Barton, R. A., Schulman, A., Jacobson, E. L., and Jacobson, M.,** Chromatographic separations of pyridine and adenine nucleotides on thin layers of poly (ethyleneimine)-cellulose, *J. Chromatogr.,* 130, 145, 1977.
39. **Raaen, H. P. and Kraus, F. E.,** Resolution of complex mixtures of nucleic acid bases, nucleosides and nucleotides by two-dimensional thin-layer chromatography on polyethyleneimine-cellulose, *J. Chromatogr.,* 35, 531, 1968.
40. **Schwartz, P. M. and Drach, J. C.,** Thin-layer chromatography of purine bases, nucleosides and nucleotides, in *Nucleic Acids Chemistry,* Vol. 2, Townsend, L. B. and Tipson, S. P., Eds., John Wiley & Sons, New York, 1978, 1061.

40a. **Bochner, B. R. and Ames, B. N.,** Complete analysis of cellular nucleotides by two-dimensional thin-layer chromatography, *J. Biol. Chem.,* 257, 9759, 1982.

41. **Lee, P. C., Bochner, B. R., and Ames, B. N.,** Diadenosine 5′,5‴-P[1],P[4]-tetraphosphate and related adenylated nucleotides in *Salmonella typhimurium, J. Biol. Chem.,* 258, 6827, 1983.
42. **Bieleski, R. L.,** Separation of phosphate esters by thin-layer chromatography and electrophoresis, *Anal. Biochem.,* 12, 230, 1965.
43. **Cole, C. V. and Ross, C.,** Extraction, separation and quantitative estimation of soluble nucleotides and sugar phosphates in plant tissues, *Anal. Biochem.,* 17, 526, 1966.
44. **Kumar, D. S. S., Pinck, L., and Hirth, L.,** Separation of 2′- and 3′-mononucleotides in plant viral ribonucleic acids by an improved two-dimensional thin-layer chromatography, *Anal. Biochem.,* 48, 497, 1972.
45. **Hashimoto, S., Wold, W. S. M., Brackmann, K. H., and Green, M.,** Nucleotide sequences of 5′-termini of adenovirus 2 early transforming region Ela and Elb messenger ribonucleic acids, *Biochemistry,* 20, 6640, 1981.
46. **Van Gennip, A. H., Van Noordenburg-Huistra, D. Y., De Bree, P. K., and Wadman, S. K.,** Two-dimensional thin-layer chromatography for the screening of disorders of purine and pyrimidine metabolism, *Clin. Chim. Acta,* 86, 7, 1978.
47. **Seipel, S. and Reichert, U.,** Two-dimensional thin-layer chromatography-autoradiography of intracellular purine interconversion products, *J. Chromatogr.,* 135, 485, 1977.
48. **Pataki, G.,** Thin-layer chromatography of nucleic acid bases, nucleosides, nucleotides and related compounds. III. Separation of complex mixtures on cellulose layer, *J. Chromatogr.,* 29, 126, 1967.
49. **Munns, T. W., Podratz, K. C., and Katzman, P. A.,** Separation of methylated bases of ribonucleic acids by two-dimensional thin-layer chromatography, *J. Chromatogr.,* 76, 401, 1973.
50. **Munns, T. W. and Sims, H. F.,** Separation of the methylated nucleoside constituants by two-dimensional thin-layer chromatography, *J. Chromatogr.,* 111, 403, 1975.
51. **Das, I.,** Separation of picomolar concentrations of adenine and guanine nucleotides and nucleosides by high performance thin-layer chromatography on silica gel, *Biochem. Soc. Trans. 576th Meet. (London),* 6, 1047, 1978.
52. **Mirzabekov, A. D. and Griffin, B. E.,** 5s RNA conformation. Studies of its partial T1 ribonuclease digestion by gel electrophoresis and two-dimensional thin-layer chromatography, *J. Mol. Biol.,* 72, 633, 1972.
53. **Chen, E. Y. and Roe, B. A.,** A two-dimensional thin-layer chromatographic procedure for the sequential analysis of oligonucleotides employing tritium post-labeling, *Nucleic Acids Res.,* 4, 3563, 1977.
54. **Miller, J. S. and Burgess, R. R.,** Selectivity of RNA chain initiation *in vitro.* I. Analysis of RNA initiations by two-dimensional thin layer chromatography of 5′-triphosphate labeled oligonucleotides, *Biochemistry,* 17, 2055, 1978.
55. **Gross, H. J., Kroath, H., Janda, H. G., and Jungwirth, C.,** Analysis of the methylated "cap" structures of Vaccinia mRNA by two-dimensional thin-layer chromatography, *Mol. Biol. Rep.,* 4, 105, 1978.
56. **Lando, D., de Rudder, J., and Privat de Garilhe, M.,** Separation of apurinic DNA oligonucleotides by two-dimensional thin-layer chromatography on silica gel, *J. Chromatogr.,* 30, 143, 1967.
57. **Bergquist, P. L.,** A thin layer method for the mapping of enzymatic digest of ribonucleic acids, *J. Chromatogr.,* 19, 615, 1965.
58. **Mundry, K. W. and Priess, H.,** Quantitative techniques for mapping oligonucleotides on thin layers of cellulose, *Biochem. Biophys. Acta,* 269, 225, 1972.
59. **Diels, D. and de Wachter, R.,** An RNA sequencing method based on a two-dimensional combination of gel electrophoresis and thin layer chromatography, *Arch. Int. Physiol. Biochim.,* 88, 26, 1980.

Section VI

HPLC OF PHARMACOLOGICALLY IMPORTANT NUCLEOBASE AND NUCLEOSIDE ANALOGS: 6-THIOGUANINE, 5-FLUOROURACIL, AND CYTARABINE

Paul A. Andrews

INTRODUCTION

The maturation and acceptance of HPLC as a routine analytical tool in the past decade have had a major impact on pharmacologic research. One area of pharmacology that has benefited greatly from this technology is the study of therapeutically useful nucleobase and nucleoside analogs such as azathioprine, ftorafar, 5-fluorouracil, 5-fluoro-2′-deoxyuridine, 6-thioguanine, cytarabine, 6-mercaptopurine, 3-deazauridine, and 5-azacytidine. HPLC-based techniques have contributed tremendously to knowledge about these drugs at the molecular, cellular, and clinical levels of research. The understanding of the metabolism and pharmacokinetics of these drugs has become an essential component of their effective clinical utilization, and drug-specific HPLC assays have greatly augmented this understanding.[1] In addition, these drug-specific HPLC assays, together with HPLC assays for monitoring changes in endogenous nucleobase, nucleoside, nucleotide, and deoxynucleotide pools (as described in other sections of this handbook), have played an important role in unraveling the mechanism of action of these nucleobase and nucleoside analogs at the cellular and molecular level. It will be the purposes of this review to document the drug-specific HPLC methods that have played a role in the research on three antimetabolites of central importance in antineoplastic chemotherapy. These drugs are 6-thioguanine (6TG), 5-fluorouracil (5FU), and cytarabine (araC).

6-THIOGUANINE

Background

6TG, first synthesized by Elion and Hitchings[2] in 1955, is used principally in the treatment of acute leukemias.[3] The analytical techniques devised over the years for the study of the pharmacology of this drug have included chromatographic separations on Dowex®-50 columns with spectrophotometric quantitation,[4] determination of ^{14}C-labeled drug,[5] use of ^{35}S-labeled drug, and paper chromatographic separations of metabolites,[6] microbiological assays,[7] DEAE-Sephadex chromatography,[8] and spectrofluorometric assays of oxidized 6TG.[9-11] Present-day HPLC-based methodologies are now the methods of choice for determination of 6TG and its metabolites. These methods are summarized in Table 1 and are described briefly below.

The HPLC method that one chooses for pharmacologic studies of 6TG will depend intimately upon the information to be garnered. 6TG must be anabolized intracellularly to 6-thioguanosine monophosphate by the enzyme hypoxanthine-guanine phosphoribosyltransferase (HGPRT) in order to elicit its cytotoxic effects.[12,13] 6TG-monophosphate inhibits *de novo* purine synthesis at the level of phosphoribosylpyrophosphate amidotransferase. 6TG-monophosphate also inhibits inosinate dehydrogenase and guanylate kinase. 6TG-monophosphate may be further anabolized to 6TG-triphosphate which is then incorporated into RNA and into DNA, presumably, as the deoxynucleotide. The incorporation of 6TG into DNA is believed to be at the basis of its cytotoxicity, although the actual consequences of this event are unclear. 6TG and 6TG nucleotides may be methylated at the C-6 sulfur by thiopurine methyltransferases,[14,15] and these metabolites are also inhibitors of the various purine pathways. 6TG is catabolized by guanase to 6-thioxanthine, which is catabo-

Table 1
HPLC ANALYSIS OF 6-THIOGUANINE

Sample	Extraction	Column (mm × cm)	Stationary phase (μm)[a]	Mobile phase	Flow rate (mℓ/min)	Retention time (min)	Detection	Sensitivity[b]	Ref.
Cells	PCA/KOH	1 × 300	Pellicular anion-exchange[c] (NR[d])	0.015—0.25 *M* KH_2PO_4, 2.2 *M* KCl	0.2	9	340 nm, ^{35}S	5 nmol/mℓ	8, 17—19
	PCA/KOH	4.6 × 25	Partisil®SAX[e]	0.01—1.0 *M* $NaPO_4$[f], pH 3.5	1.8	—	340 nm	—	16
	PCA/KOH	4.6 × 25	BP-AN6[g] (NR)	0.5 *M* NH_4COOH, pH 4.6/ 5% 2-propanol	1.2	15	340 nm	—	16
	PCA, phosphatase	1.8 × 40	M71[h]	0.4 *M* NH_4CHOO, pH 4.6	0.22	16	342 nm		20, 21
Cells, plasma	(Cells) PCA/KOH, oxidation; (plasma) ultrafiltration, oxidation	4.6 × 25	Partisil® SAX	5 m*M* KPO_4, pH 3.5—0.25 *M* KPO_4, pH 4.5, 0.5 *M* KCl	5.0	7	Fluorescence Ex: 330 nm Em: >389 nm	40 pmol/mℓ	22
Plasma	PCA/KOH	4.6 × 25	Partisil® SAX	15 m*M* KPO_4, pH 4.0	1.3	12	Fluorescence	—	23
Plasma, urine	Ultrafiltration	4 × 15	Nucleosil® $5C_8$[i](5)	60 m*M* Na_2HPO_4, 6.3 m*M* citric acid, pH 7.0	1.4	5.7	343 nm	1.2 nmol/mℓ	24
Plasma	PCA/KOH	3.9 × 30	µBondapak® C_{18}[j]	10 m*M* Na-acetate, pH 3.5/ 10% methanol	2.0	4.0	340 nm	0.8 nmol/mℓ	25, 26
				10 mM Na-acetate, pH 3.5/ 25% methanol			310 nm		

[a] 10 μm particle size unless indicated.
[b] By external standard quantitation.
[c] Varian Corp., Palo Alto, Calif.
[d] Not reported.
[e] Whatman, Clifton, N.J.
[f] Sodium phosphate buffer, KPO_4 indicates a potassium phosphate buffer.
[g] Benson, Co., Reno, Nev.
[h] Beckman, Munich, West Germany
[i] Macherey-Nagel, Duren, West Germany
[j] Waters Associates, Milford, Mass.

lized by xanthine oxidase to 6-thiouric acid. 6TG-monophosphate can be inactivated to 6-thioguanosine by alkaline phosphatase. 6-Thioguanosine can also be generated directly from 6TG by the action of purine nucleoside phosphorylase.[16] Thus, it is clear that the anabolic and catabolic metabolism of 6TG generates a broad range of products. The metabolites of specific experimental interest will dictate the choice of HPLC method.

Ion-Exchange HPLC

The first HPLC method, described by Brown[17] and used by others[8,18,19] for the analysis of 6TG and metabolites, was somewhat primitive and reflected the state of the art at the time. Nonetheless, this method provided the basis for a number of excellent studies regarding the biochemical pharmacology of 6-mercaptopurine and 6TG. This method separated 6TG, 6TG-monophosphate, and 6-thiouric acid (among other thiopurines) on a long column of pellicular anion-exchange resin followed by detection of either radioactivity or UV absorbance. Biochemical and cellular pharmacologic studies of 6TG are primarily interested in determining the levels of 6TG and its active metabolite 6TG-monophosphate. Strong anion-exchange HPLC assay of 6TG and its nucleotide metabolites is thus an ideal choice of methodology for these studies. An updated, direct, strong anion-exchange method using a microparticulate stationary phase and a conventional column size has been briefly reported by Lee and Sartorelli.[16] Cation-exchange HPLC has been used to separate a large number of 6TG and 6-mercaptopurine metabolites in cell extracts.[16,20,21] These cation-exchange HPLC methods, however, either ignore thiopurine metabolites at the nucleotide level or estimate their levels indirectly after conversion to nucleosides by treatment of cell lysates with alkaline phosphatase.

Tidd and Dedhar have described a method that combines the sensitivity of spectrofluorometric determinations of 6TG with the specificity of HPLC.[22] This method has been used both for cellular pharmacology studies[22] and to study the pharmacokinetics of orally administered 6TG in man.[23] Perchloric acid (PCA) soluble extracts of cells or plasma ultrafiltrates are treated with potassium permanganate which oxidizes the 6-thiopurines to the fluorescent 6-sulfonopurines. These compounds are then separated on a strong anion-exchange column. Because of the high sensitivity of fluorescent detection, this method is useful for cellular biochemical studies and pharmacokinetic studies of oral or constant infusion 6TG, where the levels to be measured are in the submicromolar range. However, this method suffers from the disadvantage that the methylated metabolites of 6TG along an important catabolic pathway are not measured since they are not oxidized.

Reversed-Phase HPLC

Recently, two reversed-phase HPLC methods for the assay of 6TG and metabolites have been described.[24,25] These methods offer similar sensitivities around 1 nmol/mℓ and rapid separation of 6TG and metabolites. Thus, they are ideally suited for pharmacokinetic studies. Typical chromatograms using the method of Andrews et al.[25] are shown in Figure 1. Two separate mobile phases and detection at two separate wavelengths were used to determine the full spectrum of 6-thio and methylated metabolites. Adaptation of these conditions to gradient elution and dual wavelength detection should expedite the assay further. The method of Andrews et al. has been applied to the study of high-dose i.v. 6TG pharmacokinetics in humans.[26] This study found that 6TG given in high doses was metabolized to a large degree to methylated metabolites (Figure 2). One of these metabolites was identified as 6-methylthioguanosine by mass spectrometry and two others remained unidentified.

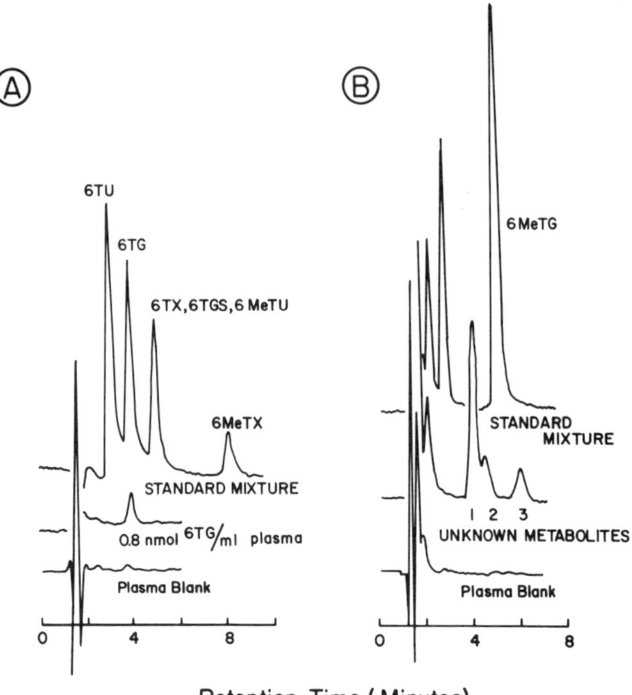

FIGURE 1. Chromatograms of 6TG and metabolites using two isocratic systems. (A) Elution with 10 m*M* Na-acetate, pH 3.5/10% methanol and detection at 340 nm. The traces show a plasma blank, the detection limit, and a mixture of 6TG metabolites used as standards; (B) elution with 10 m*M* Na-acetate pH 3.5/25% methanol and detection at 310 nm. The traces show a plasma blank; a typical patient's plasma extract showing two unknown methylated metabolites and 6-methylthioguanosine (unknown 3); and a mixture of metabolite standards.[25] (From Andrews, P. A., Egorin, M. J., May, M. E., and Bachur, N. R., *J. Chromatogr.*, 227, 83, 1982. With permission.)

5-FLUOROURACIL

Background

5FU is an effective palliative agent against a variety of solid tumors and is curative in multiple basal cell carcinomas. As with 6TG, it must be metabolized to the nucleotide level to exert its cytotoxic effect. 5FU is extensively metabolized by pyrimidine biosynthetic and degradative pathways as described elsewhere,[3,27-29] and the specific metabolic steps will not be reiterated here. The products to be expected from the anabolic metabolism of 5FU include 5-fluorouridine (FUR) and its mono-, di-, and triphosphates (FUMP, FUDP, and FUTP), and 5-fluoro-2′-deoxyuridine (FUdR) and its mono- and diphosphates (FdUMP and FdUDP). 5FU is catabolized, principally by the liver, to dihydro-5FU which is sequentially cleaved to α-fluoro-β-ureidopropionic acid; α-fluoro-β-guanidopropionic acid; and α-fluoro-β-alanine, CO_2, and ammonia. FUDP-hexoses have recently been identified in L1210 cells in significant quantities, although their role in 5FU cytotoxicity is unknown.[30] The cytotoxicity of 5FU can be attributed to two major effects: irreversible inhibition of thymidylate synthetase by FdUMP, and the incorporation of FUTP into RNA with subsequent inhibition of RNA processing and function.

Numerous historical methods for 5FU determination such as thin-layer chromatography

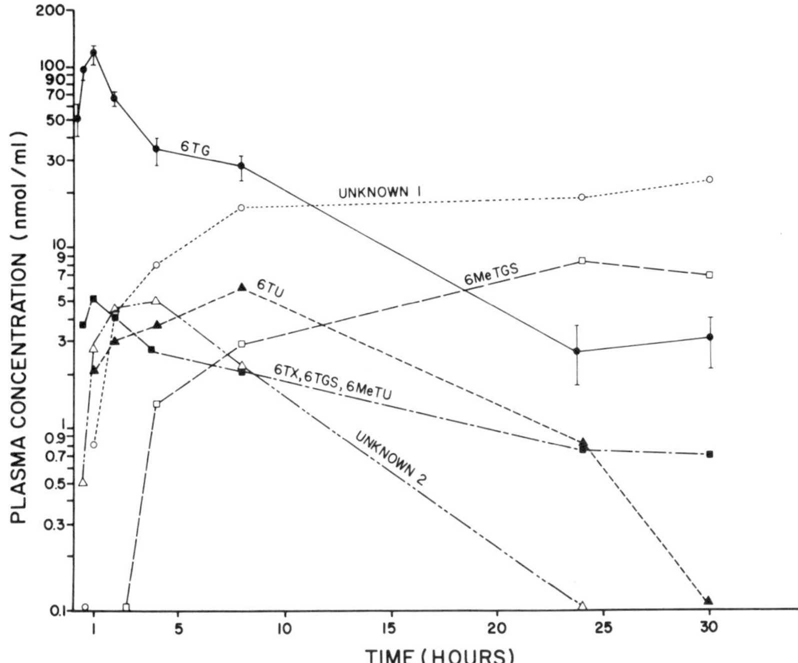

FIGURE 2. Plasma concentrations of 6TG and metabolites in a patient receiving 1800 mg 6TG as a constant 60-min infusion. Points represent the mean of triplicate extractions. Error bars for 6TG indicate ± the standard deviation.[25] (From Andrews, P. A., Egorin, M. J., May, M. E., and Bachur, N. R., *J. Chromatogr.*, 227, 83, 1982. With permission.)

(TLC), fluorine analysis, volumetric analysis, and spectrophotometric analysis are documented elsewhere along with physicochemical properties of 5FU.[31] Conventional column chromatography with Dowex®-1 anion-exchange resin was used originally to identify the intracellular metabolites in Ehrlich ascites cells.[32] A variety of gas chromatographic (GC) methods have been used for the assay of 5FU.[33] When coupled to mass spectrometric detectors, these (GC) methods offer very high sensitivity and the ability to routinely determine the major plasma metabolite dihydro-5FU.[34] A number of HPLC assays of 5FU and related fluorinated nucleosides and nucleotides have recently come to the forefront as the most useful methods for the determination of 5FU and its nucleoside and nucleotide metabolites in plasma, cells, and tissues. These methods are summarized in Table 2 and are described below.

Reversed-Phase HPLC

As reasoned above for 6TG, the HPLC method chosen for the pharmacologic study of 5FU will again depend on the specific information to be obtained. If the primary interest is in the determination of the parent 5FU itself, as for pharmacokinetic studies, then reversed-phase HPLC is sufficient to accomplish this task. A number of methods based on C_{18} stationary phases and phosphate-buffered mobile phases have been described.[35-43] The phosphate buffer concentrations range from 2 to 50 m*M* and the pH values range from to 4.0 to 5.7. Neither of these parameters have any significant effect on 5FU retention.[37,58,60] Na-acetate-buffered mobile phases with small percentages of methanol have also been used to elute 5FU from C_{18} columns.[44-46] Water alone was sufficient to elute 5FU[30,47] and this HPLC method was used to study the effects of 5FU on RNA metabolism and thymidylate synthetase inhibition in Novikoff hepatoma cells.[48] Diasio and Wilburn[42] studied the effect of column

Table 2
HPLC ANALYSIS OF 5-FLUOROURACIL

Sample	Extraction	Column (mm × cm)	Stationary phase (μm)[a]	Mobile phase	Flow rate	Retention time	Detection	Sensitivity	Quantitation	Ref.
Plasma	Dowex® AG 1-X2, ethyl acetate	7.8 × 30	μBondapak® C_{18}[b]	10 m*M* KPO_4[c], pH 4.0	2.0	12.5	280 nm	50 pmol/mℓ	[^{3}H]-5FU I.S.	35, 36
	PCA, Alamine®/ Freon®	3.9 × 30	μBondapak® C_{18}	50 m*M* KPO_4, pH 4.6	1.5	4.5	280	0.1 nmol/mℓ	Ex. std.	37
	Na_2SO_4, ether/*n*-propanol	3.9 × 30	μBondapak® C_{18}	50 m*M* KPO_4, pH 4.6	1.3	5.0	254	0.19 nmol/mℓ	5-Fluorocytosine I.S. (3.8)[d]	38
					1.0	1.0	266	0.3 nmol/mℓ	Br-uracil®I.S. (12.0)	39
	$(NH_4)_2SO_4$, ether/2-propanol	3.9 × 30	μBondapak® C_{18}	10 m*M* KPO_4, pH 5.5 stepped to 8% methanol	1.2	6.0	265	0.77 nmol/mℓ	Thymidine ex. std. (22)	40
	$(NH_4)_2SO_4$, ether/2-propanol	4.6 × 25	Partisil® ODS-2[e]	10 m*M* KPO_4, pH 5.5	2.0	5.5	265	0.77 nmol/mℓ	Uridine ex std. (10.5)	41
Standards	—	4.0 × 25	RP-18[f]	2 m*M* $NaPO_4$, pH 5.7	0.5	13.9	254	—	—	42
Cells	TCA						[^{3}H]	—	Specific activity	43
	Ethyl acetate	3.9 × 30	μBondapak® C_{18}	10 m*M* Na-acetate, pH 4.0/15% methanol	2.0	2.0	280	0.77 nmol/mℓ	Ex. std.	44
Serum	Ethyl acetate	3.9 × 30	μBondapak® C_{18}	10 m*M* Na-acetate, pH 4.0/5% methanol	2.0	2.5	270	0.19 nmol/mℓ	Ex. std.	45, 46
Plasma	Ultrafiltration	3.9 × 30	μBondapak® C_{18}	H_2O	2.0	3.3	270	—	Ex. Std.	47, 48
Serum	PCA/KOH	3.9 × 30	μBondapak® CN[b]	5 m*M* NH_4-acetate	1.0	2.7	254	5 nmol/mℓ	Br-deoxyuridine I.S.(33)	49, 50
Plasma	Ethyl acetate/2-propanol	4.6 × 25	Partisil® SCX[e]	14 m*M* NH_4PO_4, pH 3.5	1.0	3.6	266	0.39 nmol/mℓ	Chlorpheniramine I.S. (2.2)	51, 52
	$(NH_4)_2SO_4$, ether/*n*-propanol	2 × 30	Aminex® A-25[g] (17.5)	0.3 *M* acetate, pH 4.5	0.7	8.0	254	0.78 nmol/mℓ	5-Chlorouracil I.S. (17)	53, 54
	$(NH_4)_2SO_4$, $CHCl_3$ 2-propanol	3 × 50	Silica gel	2-Propanol/acetic acid/chloroform, 15/1/84	0.7	4.0	272	0.77 nmol/mℓ	Ex. std.	55
Plasma, saliva, bile	Ether/*n*-propanol	3.9 × 30	μBondapak® C_{18}	0.12 m*M* heptane-sulfonic acid in 5% acetic acid	1.5	2.0	254	1.9 nmol/mℓ	5-Fluorocytosine I.S. (4.0)	56, 57

Cells	TCA, Alamine®/ Freon®	4.6 × 25	LiChrosorb®[h] RP-18	0—40% methanol in 5 mM $(Bu)_4NHSO_4$, 5mM KPO_4, pH 7.0	2.0	6.6	[^{3}H]	—	Specific activity	30
	Sonication, heat, snake venom, TCA, alamine/Freon			H_2O	1.0	6.6				
Standards	—	3.9 × 30	µBondapak® C_{18}	0.1 mM $(Bu)_4NHSO_4$, 2.5 mM $(Et)_4NHSO_4$, 2.0 mM Na-acetate, 1.5 mM KPO_4, pH 6.0/ 2% methanol stepped to 30 mM KPO_4	—	4.0	254	—	—	58
Hepatocytes	Sonication, centrifugation	Two 4.6 × 25 in series	Hypersil® ODS (5)[f] and Spherisorb® ODS (5)[f]	0—50% methanol in 5 mM $(Bu)_4NHSO_4$, 1.5 mM KPO_4, pH 8.0	1.0	17.2	200 nm [^{3}H]	—	Specific activity	59
Serum	AG 1-X4 resin	3.2 × 15	RSIL C_{18} HL[i] (5)	0.15 mM $(Bu)_4NH_2PO_4$, 20 mM KPO_4, pH 5.0/5% methanol	0.8	1.8	270	0.77 nmol/mℓ	1-(2-Deoxy-β-D-lyxofuranosyl)-5FU I.S. (4.0)	60

[a] 10 µm particle size unless indicated.
[b] Waters Associates, Milford, Mass.
[c] Potassium phosphate buffer, $NaPO_4$, and NH_4PO_4 represent sodium and ammonium phosphate buffers.
[d] Retention time (min) of internal standard.
[e] Whatman, Clifton, N.J.
[f] Brownlee Labs, Santa Clara, Calif.
[g] Bio-Rad, Richmond, Calif.
[h] Altex, Berkeley, Calif.
[i] RSL, St. Martens-Latem, Belgium.

temperature on the resolution of 5FU and its anabolites and found that subambient column temperatures (5°C) lead to considerable retention and the resolution of FUMP and FdUMP by reversed-phase HPLC, but at the sacrifice of reduced sensitivity. This method (25°C) was applied to the detailed analysis of FUdR membrane transport and metabolism in Ehrlich ascites tumor cells.[43] A separation of 5FU on a cyano stationary phase operating in the reversed-phase mode has been described and applied to the measurement of 5FU in serum.[49,50] These reversed-phase assays are also applicable to the determination of the nucleoside metabolites of 5FU (FUR and FUdR) and the analog ftorafur.

Ion-Exchange HPLC

Ion-exchange stationary phases have been developed to quantify 5FU in complex biological matrices. Strong cation-exchange HPLC was used to measure 5FU in human plasma[51] and to study the uptake of 5FU in rat liver slices.[52] This method obviously will not determine nucleotide metabolites of 5FU. A strong anion-exchange HPLC method was reported for the assay of 5FU in plasma.[53] An adaptation of this method was utilized to study 5FU metabolism in tissue culture cells, however, FUR and FUdR are not resolved and the di- and triphosphates are eluted as a single peak.[54]

Normal Phase HPLC

A single normal phase HPLC method has been reported and applied to the pharmacokinetic study of 5FU in man.[55]

Ion-Pair HPLC

Both sulfonate and ammonium ion-pairing reagents have proved useful for the analysis of 5FU. A modification of the method of Christophidas et al.[38] that uses a heptanesulfonic acid mobile phase was applied to the analysis of 5FU in plasma and saliva.[56,57] Perhaps the best choice of HPLC methodology for investigating the cellular pharmacology of 5FU is method that incorporates a tetrabutylammonium counter-ion in the mobile phase. These methods can separate and quantify a large number of 5FU metabolites including FUR, FUdR, FUDP, FUTP, FdUMP, FdUDP, FUDP-hexoses, and 5FU catabolites. Au et al.[58] have reported an in-depth study of the effect of stationary phases, pH, buffer concentration, and counter-ion chain length and concentration on the retention of 5FU and its metabolites (or nonfluorinated analogs). The optimal conditions for separating these compounds were a complex mobile phase and a two-step elution program as described in Table 2. Sommadossi et al.[59] employed a tetrabutylammonium sulfate counter-ion, methanol gradient, and two columns in series to separate 5FU, its anabolites, and its catabolites (dihydro-5FU, α-fluoro-β-ureidopropionic acid, and α-fluoro-β-alanine). This method was applied to the investigation of 5FU metabolism in isolated hepatocytes. The rapid catabolism and complete lack of anabolites in these cells emphasizes the critical role that the liver plays in 5FU clearance. Gelijkens and Leenheer reported an HPLC method with a tetrabutylammonium counter-ion and also studied the influence of buffer concentration, counter-ion concentration, and mobile phase pH on the retention of 5FU and its anabolites. This method was applied to the determination of FUdR in plasma.[60] Pogolotti et al.[30] reported an integrated methodology that permits the complete analysis of the intracellular anabolites of 5FU. The primary separation of metabolites was performed with a mobile phase containing a tetrabutylammonium sulfate counter-ion and a methanol gradient which separated the nucleosides, nucleotides, and hexoses of 5FU. Since FUR and FUdR were eluted as a fused peak, more accurate quantification of these metabolites was obtained by chromatography by a reversed-phase method.[30] In addition, procedures for the determination of the FdUMP covalently bound to thymidylate synthetase were described.

Detection and Extraction

All these methods either use a variety of wavelengths between 254 and 280 nm or the radioactivity of 6-[^{3}H]-5FU for detection of peaks. The detection limits of UV absorbance are approximately 0.5 nmol/mℓ for 0.5 to 1.0 mℓ of plasma. Buckpitt and Boyd[35] report a detection limit of 50 pmol/mℓ; however, they must extract 5 to 10 mℓ of plasma to measure this low level. Use of 280 nm seems to give less interference from background peaks in plasma, particularly when PCA extracts are used.[37] The [^{3}H]-labeled drug is essential for studying the cellular pharmacology of 5FU, i.e., detecting the low levels of anabolites and catabolites found in cells.

A variety of extraction methods have been used for 5FU analysis in plasma or serum; the most widely applied is extraction with six to ten volumes of ether/2-propanol or ether/*n*-propanol, 21/4 (v/v). This gives recoveries from plasma in the range of 70 to 80%. Hsu and Marrs[51] conducted a detailed analysis of various solvent combinations for the extraction of 5FU from water and found ethyl acetate/2-propanol, 21/4 (v/v) to be the most efficient (77.8%). Sodium or ammonium sulfate are often used to salt out the aqueous phase in these extractions, however, these salts did not increase the extraction efficiency with ethyl acetate/2-propanol[51] and affected the extraction of the internal standard 5-fluorocytosine differently than 5FU.[39] Redissolving dried residues or back extracting organic extracts into basic solvents takes advantage of the weak acidity of 5FU (pK_a = 8.0 [33]) and increases recovery of 5FU while recovery of interfering peaks is decreased.[38,46,53] Ethyl acetate, PCA, trichloroacetic acid (TCA), and ion-exchange-resin-based extractions have also been used for 5FU analysis in biological samples. A number of internal standards have been utilized as shown in Table 2.

CYTARABINE

Background

AraC (1-β-D-arabinofuranosylcytosine) is a pyrimidine deoxynucleoside analog clinically useful in the treatment of hematologic malignancies (especially acute myelogenous leukemia) and, as recently reported, the treatment of ovarian carcinoma.[61,62] AraC must be intracellularly converted by a series of kinases to its triphosphate, araCTP, to exert its cytotoxicity.[3,13,63,64] AraCTP possesses multiple site of action, but the major mechanism of cytotoxicity can be attributed to its potent inhibition of DNA polymerases and, hence, DNA synthesis. The exact molecular basis of this inhibition, whether competition with dCTP or inhibition of chain initiation or elongation after incorporation into nascent DNA strands, remains more controversial.[63,65-65b] AraC is rapidly inactivated to uracil arabinoside (araU) by cytidine deaminase which is found in plasma, liver, kidney, the gastrointestinal tract, and some tumor cells. AraCMP can be deaminated by dCMP deaminase to generate araUMP.

The determination of araC in biological fluids is a difficult problem due to the close similarity of its structure and chemical properties to the endogenous nucleosides cytidine and deoxycytidine. Before the advent of HPLC and its application to the analysis of this drug, araC was determined by a variety of techniques such as paper chromatography,[66,67] column chromatography,[68] microbiological assays,[7,69-71] bioassays,[72,73] TLC,[74,75] radio-enzymatic assay,[75] gas chromatography/mass spectrometry,[76-78] or RIA.[79,80] HPLC methods have replaced many of these techniques, with the exception of RIA, for studying the cellular and clinical pharmacology of araC. Since araC is an S-phase-specific agent and low levels are adequate for activity, therapeutic plasma levels are often below the limit of detection of UV absorbance detectors in HPLC. RIA has thus remained a popular method for determining these low levels in plasma with a detection limit of 2 to 4 pmol/mℓ.[79,86] These HPLC methods are summarized in Table 3 and described briefly below.

Table 3
HPLC ANALYSIS OF CYTARABINE AND CYTARABINE-TRIPHOSPHATE

Sample	Extraction	Column (mm × cm)	Stationary phase (μm)[a]	Mobile phase	Flow rate	Retention time	Detection	Sensitivity[b]	Ref.
Plasma	Ultrafiltration PCA/ K_2HPO_4	4.6 × 50 (two columns)	Nucleosil® C_{18}[c]	0.2 *M* KPO_4[d], pH 2.0	—	15.0	280 nm	8.2 pmol/mℓ	81
Plasma, CSF	None	4 × 30	Spherisorb® ODS[e] (5)	50 m*M* $NaPO_4$, pH 7.0	1.6	6.0	270	0.2 nmol/mℓ	82
Plasma, urine	Ultrafiltration, Affi-gel® 601	3.9 × 30	μBondapak® C_{18}[f]	10 m*M* KPO_4, pH 5.6	0.6	10.0	270	1.3 pmol	83
Plasma	$ZnSO_4$, $Ba(OH)_2$	4.6 × 25	Ultrasphere®-Octyl (5)[g]	10 m*M* KPO_4, pH 7.0/1.5% methanol	1.6	4.5	281	1.0 nmol/mℓ	84
CSF								0.5 nmol/mℓ	
Plasma, cells	PCA, KOH	4.6 × 15	LDC ODS[h] (NR[i])	2.5 m*M* KPO_4, pH 7.0/0.75% acetonitrile	1.0	5.7	254, 280	—	85, 86
Plasma, urine	Ultrafiltration	4 × 50	Aminex® A-27[j] (13.5)	25 m*M* Na-citrate, 80 m*M* Na-tetraborate, pH 9.3	0.7	14.5	270	10.3 pmol	83
DNA	Enzymatic digestion	—	MicroPak® AX-10[k] (NR)	10 m*M* KPO_4, pH 2.85/80% acetonitrile	—	5.0	[^3H]	—	65a
Cells	PCA, KOH	0.64 × 30	μBondapak® NH_2[f]	2 m*M* KPO_4, pH 2.1—0.4 *M* KPO_4, pH 5.0	0.3	0.9	254	—	87
Plasma	TCA, NH_4CHOO	4.6 × 25	Partisil® SCX[l]	0.01 *M* NH_4COOH	1.0	11.0	254	82 pmol/mℓ	88
	—	— × 13	Aminex® A-7[j] (NR)	20 m*M* KPO_4, pH 4.3	—	12	254	—	89
	PCA/KOH	8 × 10	μBondapak® C_{18}	0—25% acetonitrile in 10 m*M* heptane-sulfonic acid, pH 3.5	2.0	9.1	280	0.5 nmol/mℓ	90
Plasma, CSF, urine	Ultrafiltration	4.6 × 15 + 4.6 × 25	Ultrashpere® ODS (5)[g] + Partisil® SCX	2.5 m*M* KPO_4, pH 3.2/2.5% methanol	0.8	25.7	280	82 pmol/mℓ[m]	91
Cells	PCA	4.6 × 25	Partisil® SAX[l]	0.27—0.75 *M* NH_4PO_4, pH 3.7	3.0	16.1 (araCTP)	280	50 pmol/mℓ	92, 93
	Methanol	4 × 20	Aminex® A-29[j] (9.0)	0.25 *M* Na-citrate, pH 8.2	0.7	19.0 (araCTP)	270	4.1 nmol/mℓ	83

	PCA, KOH/ periodate	4.6 × 25	Partisil® SAX	0.5 *M* NH_4PO_4, pH 3.5/2% methanol	0.5	28.6 (araCTP)	254, 280	—	94
				0.15 *M* NH_4PO_4, pH 3.5 stepped to 0.75 M/5% methanol	0.5	52.0 (araCTP)	254, 280	—	
Myeloblasts	PCA	2.1 × 100	ABX-Permaphase®[n] (NR)	2.5 m*M* KPO_4, pH 3—0.5 *M* KPO_4, pH 4.4	0.7	25 (araCTP)	280, [^{3}H]	—	86, 95
Cells	Ethanol	4.6 × 25	Partisil® SAX	0.125 *M* KPO_4, 0.075 *M* Na_3-citrate, pH 4.6	2.5	8.0 (araCTP)	280	0.2 nmol/mℓ	96
	TCA, Dowex® AG1-X8	2.7 × 120 (two columns)	ABX-Permaphase®	5 m*M* KPO_4, pH 3.0—0.2 *M* KPO_4	3.0	7.5 (araCTP)	254, [^{3}H]	—	74

[a] 10 μm particle size unless indicated.
[b] By external standard quantitation unless indicated.
[c] Rainen Instrument Co., Woburn, Mass.
[d] Potassium phosphate buffer, $NaPO_4$, and NH_4PO_4 represent sodium and ammonium phosphate buffers.
[e] Latek, Heidelberg, F.D.R.
[f] Waters Associates, Milford, Mass.
[g] Altex, Berkeley, Calif.
[h] Laboratory Data Control, Riveria Beach, Fla.
[i] Not reported.
[j] Bio-rad, Richmond, Calif.
[k] Varian, Palo Alto, Calif.
[l] Whatman, Clifton, N.J.
[m] Adenine arabinoside internal standard.
[r] Dupont, Wilmington, Del.

Reversed-Phase HPLC

Reversed-phase HPLC methods for the determination of araC and its only metabolite, araU, have been described. All of these methods employ a phosphate-buffered mobile phase of varying concentration and pH.[81-86] Three of these methods use C_{18} stationary phases bonded to spherical silica particles.[81,82,84] Two approaches have been taken to the elimination of interferences from endogenous nucleosides on these reversed-phase columns. One can use either high efficiency columns to separate these compounds from araC (provided by 5-μm particle sizes[82,84] or two 10 μm columns in series[81]), or treatment of extracts with boronate columns[83] or periodate oxidation[94] to remove ribonucleosides prior to chromatography. A number of these methods have been applied to clinical pharmacokinetic studies. King et al.[61] used the method of Breithaupt and Schick[82] to study the pharmacokinetics of araC in the plasma and peritoneal fluid after instillation of araC into the peritoneal cavity. Zimm et al.[84] used reversed-phase HPLC to study the pharmacokinetics of araC in plasma and cerebrospinal fluid after intrathecal administration.

Ion-Exchange HPLC

Methods using both weak and strong anion-exchange columns have been reported.[65a,83,87] Strong cation-exchange columns can also be used,[88,89] although araC is poorly resolved from plasma constituents eluted at the void volume.[88] Sinkule and Evans have described a dual-column method that places a C_{18} column in series with a strong anion-exchange column to measure araC in plasma.[91] Cytidine and deoxycytidine are well separated from araC with this method.

Ion-Pair HPLC

An ion-pair method using heptanesulfonic acid in the mobile phase has recently been reported by Murphy et al.[90] AraC was highly retained and required an acetonitrile gradient for its elution.

Extraction

An essential feature of determining araC in plasma is to draw blood into tubes containing the cytidine deaminase inhibitor tetrahydrouridine (usually at a final concentration of 0.1 m*M*) to prevent deamination during sample storage or work-up.[81,82,88,90,91] PCA or ultrafiltration has been the most useful means of extraction. PCA was found to be superior to TCA and filtration for extracting araCTP from cell homogenates.[87] Ethanol is useful for extracting small sample volumes (cells) and does not require a neutralization step.[96] Adenine arabinoside and 5-fluorodeoxyuridine are useful internal standards.[90,91] One method injects samples of plasma directly onto the column, however, the columns were only good for up to 4 mℓ of total plasma injected.[82] Extraction of plasma with organic solvents gives poor recovery of araC and is not a viable method.[82]

Cytarabine Triphosphate

Since the major determinant of araC cytotoxicity is the formation of araCTP, there has been much interest in measuring intracellular levels of this nucleotide in tumor cells from treated patients as a means of monitoring therapy and indicating the prospective success of chemotherapy.[86,92,93,96] A number of methods designed specifically for determining araCTP are, therefore, also listed in Table 3. All of these methods employ strong anion-exchange columns.[74,83,86,92-96] It will be of interest for future development to study whether ion-pair techniques similar to those used for the simultaneous assay of 5FU and its nucleotide metabolites can be applied to simultaneous assay of araC and araCTP. This will prove very useful for the study of the cellular pharmacology of araC.

ACKNOWLEDGMENT

I thank Sharon Don for skilled assistance in the preparation of this manuscript.

REFERENCES

1. **Erlichman, C., Donehower, R. C., and Chabner, B. A.,** The practical benefits of pharmacokinetics in the use of antineoplatic agents, *Cancer Chemother. Pharmacol.*, 4, 139, 1980.
2. **Elion, G. B. and Hitchings, G. H.,** The synthesis of 6-thioguanine, *J. Am. Chem. Soc.*, 77, 1676, 1955.
3. **Chabner, B. A. and Meyers, C. E.,** Clinical pharmacology of cancer chemotherapy, in *Cancer: Principles and Practice,* DeVita, V. T., Hellman, S., and Rosenberg, S. A., Eds., Lippincott, Philadelphia, 1982, chap. 9.
4. **Moore, E. C. and LePage, G. A.,** The metabolism of 6-thioguanine in normal and neoplastic tissues, *Cancer Res.*, 18, 1075, 1958.
5. **LePage, G. A. and Jones, M.,** Further studies on the mechanism of action of 6-thioguanine, *Cancer Res.*, 21, 1590, 1961.
6. **LePage, G. A. and Whitecar, J. P., Jr.,** Pharmacology of 6-thioguanine in man, *Cancer Res.*, 31, 1627, 1971.
7. **Pittillo, R. F. and Woolley, C.,** Disposition of arabinosylcytosine (NSC-63878) and 6-thioguanine (NSC-752) in solid L1210 leukemia tumor-bearing mice, *Cancer Chemother. Rep.*, 57, 275, 1973.
8. **Nelson, D. J., Bugge, C. J. L., Krasny, H. C., and Zimmerman, T. P.,** Separation of 6-thiopurine derivatives on DEAE-Sephadex columns and in the high-pressure liquid chromatograph, *J. Chromatogr.*, 77, 181, 1973.
9. **Finkel, J. M.,** Fluorometric assay of thioguanine, *J. Pharm. Sci.*, 64, 121, 1975.
10. **Dooley, T. and Maddocks, J. L.,** Assay of 6-thioguanine in human plasma, *Br. J. Clin. Pharmacol.*, 9, 77, 1980.
11. **Thomas, A. D.,** Spectrofluorometric determination of thiopurines-I 6-thioguanine, *Talanta,* 23, 383, 1976.
12. **Paterson, A. R. P. and Tidd, D. M.,** 6-Thiopurines, in *Handbook of Experimental Pharmacology: Antineoplastic and Immunosuppressive Agents,* Vol. 38, Part 2, Sartorelli, A. C. and Johns, D. G., Eds., Springer-Verlag, New York, 1975, chap. 48.
13. **Pratt, W. B. and Ruddon, R. W.,** *The Anticancer Drugs,* Oxford, New York, 1979, chap. 6.
14. **Woodson, L. C. and Weinshilboum, R. M.,** Human kidney thiopurine methyltransferase: purification and biochemical properties, *Biochem. Pharmacol.*, 32, 819, 1984.
15. **Woodson, L. C., Ames, M. M., Selassie, C. D., Hansch, C., and Weinshilboum, R. M.,** Thiopurine methyltransferase: aromatic thiol substrates and inhibition by benzoic acid derivatives, *Mol. Pharmacol.*, 24, 471, 1983.
16. **Lee, S. H. and Sartorelli, A. C.,** Conversion of 6-thioguanine to the nucleoside level by purine nucleoside phosphorylase of sarcoma 180 and sarcoma 180/TG ascites cells, *Cancer Res.*, 41, 1086, 1981.
17. **Brown, P. R.,** The rapid separation of nucleotides in cell extracts using high-pressure liquid chromatography, *J. Chromatogr.*, 52, 257, 1970.
18. **Scholar, E. M., Brown, P. R., and Parks, R. E., Jr.,** Synergistic effect of 6-mercaptopurine and 6-methylmercaptopurine ribonucleoside on the levels of adenine nucleotides of sarcoma 180 cells, *Cancer Res.*, 32, 259, 1972.
19. **Nelson, J. A. and Parks, R. E., Jr.,** Biochemical mechanisms for the synergism between 6-thioguanine and 6-(methylmercapto)purine ribonucleoside in sarcoma 180 cells, *Cancer Res.*, 32, 2034, 1972.
20. **Breter, H.-J.,** The quantitative determination of metabolites of 6-mercaptopurine in biological materials, *Anal. Biochem.*, 80, 9, 1977.
21. **Breter, H.-J., and Zahn, R. K.,** Quantitation of intracellular metabolites of [^{35}S]-6-mercaptopurine in L5178Y cells grown in time-course incubates, *Cancer Res.*, 39, 3744, 1979.
22. **Tidd, D. M. and Dedhar, S.,** Specific and sensitive combined high-performance liquid chromatographic-flow fluorometric assay for intracellular 6-thioguanine nucleotide metabolites of 6-mercaptopurine and 6-thioguanine, *J. Chromatogr.*, 145, 237, 1978.
23. **Brox, L. W., Birkett, L., and Belch, A.,** Clinical pharmacology of oral thioguanine in acute myelogenous leukemia, *Cancer Chemother. Pharmacol.*, 6, 35, 1981.
24. **Breithaupt, H. and Goebel, G.,** Quantitative high pressure liquid chromatography of 6-thioguanine in biological fluids, *J. Chromatogr. Sci.*, 19, 1981.

25. **Andrews, P. A., Egorin, M. J., May, M. E., and Bachur, N. R.,** Reversed-phase high-performance liquid chromatography analysis of 6-thioguanine applicable to pharmacologic studies in humans, *J. Chromatogr.*, 227, 83, 1982.
26. **Konits, P. H., Egorin, M. J., Van Echo, D. A., Aisner, J., Andrews, P. A., May, M. E., Bachur, N. R., and Wiernik, P. H.,** Phase II evaluation and plasma pharmacokinetics of high-dose intravenous 6-thioguanine in patients with colorectal carcinoma, *Cancer Chemother. Pharmacol.*, 8, 199, 1982.
27. **Heidelberger, C.,** Pyrimidine and pyrimidine nucleoside antimetabolites, in *Cancer Medicine*, Holland, J. F. and Frei, E., III, Eds., Lea & Febiger, Philadelphia, 1974, 768.
28. **Heidelberger, C.,** Fluorinated pyrimidines and their nucleosides, in *Handbook of Experimental Pharmacology: Antineoplastic and Immunosuppressive Agents*, Vol. 38, Part 2, Sartorelli, A. C. and Johns, D. G., Eds., Springer-Verlag, New York, 1975, chap. 41.
29. **Sadee, W. and Wong, C. G.,** Pharmacokinetics of 5-fluorouracil: inter-relationship of biochemical kinetics in monitoring therapy, *Clin. Pharmacokinet.*, 2, 437, 1977.
30. **Pogolotti, A. L., Jr., Nolan, P. A., and Santi, D. V.,** Methods for the complete analysis of 5-fluorouracil metabolites in cell extracts, *Anal. Biochem.*, 117, 178, 1981.
31. **Rudy, B. C. and Senkowski, B. Z.,** Fluorouracil, in *Analytical Profiles of Drug Substances*, Vol. 2, Florey, K., Ed., Academic Press, New York, 1973, 221.
32. **Chaudhuri, N. K., Montag, B. J., and Heidelberger, C.,** Studies on fluorinated pyrimidines. III. The metabolism of 5-fluorouracil-2-^{14}C and 5-fluoroorotic-2-^{14}C acid *in vivo*, *Cancer Res.*, 18, 318, 1958.
33. **Gupta, R. N., Ed.,** *Handbook of Chromatography: Drugs*, CRC Press, Boca Raton, Fla., 1981, 268.
34. **Aubert, C., Sommadossi, J. P., Coassolo, P., and Cano, J. P.,** Quantitative analysis of 5-fluorouracil and 5,6-dihydrofluorouracil in plasma by gas chromatography mass spectrometry, *Biomed. Mass Spectrom.*, 9, 336, 1982.
35. **Buckpitt, A. R. and Boyd, M. R.,** A sensitive method for determination of 5-fluorouracil and 5-fluoro-2′-deoxyuridine in human plasma by high-pressure liquid chromatography, *Anal. Biochem.*, 106, 432, 1980.
36. **Jones, R. A., Buckpitt, A. R., Londer, H. H., Myers, C. E., Chabner, B. A., and Boyd, M. R.,** Potential clinical applications of a new method for quantitation of plasma levels of 5-fluorouracil and 5-fluorodeoxyuridine, *Bull. Cancer (Paris)*, 66, 75, 1979.
37. **Wung, W. E. and Howell, S. B.,** Simultaneous liquid chromatography of 5-fluorouracil, uridine, hypoxanthine, xanthine, uric acid, allopurinol, and oxipurinol in plasma, *Clin. Chem.*, 26, 1704, 1980.
38. **Christophidas, N., Mihaly, G., Vajda, F., and Louis, W.,** Comparison of liquid- and gas-liquid chromatographic assays of 5-fluorouracil in plasma, *Clin. Chem.*, 25, 83, 1979.
39. **Sampson, D. C., Fox, R. M., Tattersall, M. H. N., and Hensley, W. J.,** A rapid high-performance liquid chromatographic method for quantitation of 5-fluorouracil in plasma after continuous intravenous infusion, *Ann. Clin. Biochem.*, 19, 125, 1982.
40. **MacMillian, W. E., Wolberg, W. H., and Welling, P. G.,** Pharmacokinetics of fluorouracil in humans, *Cancer Res.*, 38, 3479, 1978.
41. **Phillips, T. A., Howell, A., Grieve, R. J., and Welling, P. G.,** Pharmacokinetics of oral and intravenous fluorouracil in humans, *J. Pharm. Sci.*, 69, 1428, 1980.
42. **Diasio, R. B. and Wilburn, M. E.,** Effect of subambient column temperature on resolution of fluorouracil metabolites in reversed-phase high performance liquid chromatography, *J. Chromatogr. Sci.*, 17, 565, 1979.
43. **Bowen, D., Diasio, R. B., and Goldman, I. D.,** Distinguishing between membrane transport and intracellular metabolism of fluorodeoxyuridine in Ehrlich ascites tumor cells by application of kinetic and high performance liquid chromatographic techniques, *J. Biol. Chem.*, 254, 5333, 1979.
44. **Wu, A. T., Au, J. L., and Sadee, W.,** Hydroxylated metabolites of R,S-1-(Tetrahydro-2-furanyl)-5-fluorouracil (ftorafur) in rats and rabbits, *Cancer Res.*, 38, 210, 1978.
45. **Hornbeck, C. L., Griffiths, J. C., Floyd, R. A., Ginther, N. C., Byfield, J. E., and Sharp, T. R.,** Serum concentrations of 5-FU, ftorafur, and a major serum metabolite following ftorafur chemotherapy, *Cancer Treat. Rep.*, 65, 69, 1981.
46. **Hornbeck, C. L., Floyd, R. A., Griffiths, J. C., and Byfield, J. E.,** Improved liquid chromatographic assay for serum fluorouracil concentrations in the presence of ftorafur, *J. Pharm. Sci.*, 70, 1163, 1981.
47. **Benvenuto, J. A., Lu, K., and Loo, T. L.,** High-pressure liquid chromatographic analysis of ftorafur and its metabolites in biological fluids, *J. Chromatogr.*, 134, 219, 1977.
48. **Cory, J. G., Breland, J. C., and Carter, G. L.,** Effect of 5-fluorouracil on RNA metabolism in Novikoff hepatoma cells, *Cancer Res.*, 39, 4905, 1979.
49. **Ensminger, W. D., Rosowsky, A., Raso, V., Levin, D. C., Glode, M., Come, S., Steele, G., and Frei, E., III,** A clinical-pharmacological evaluation of hepatic arterial infusions of 5-fluoro-2′-deoxyuridine and 5-fluorouracil, *Cancer Res.*, 38, 3784, 1978.
50. **Kirkwood, J. M., Ensminger, W., Rosowsky, A., Papthanasopoulos, N., and Frei, E.,** Comparison of pharmacokinetics of 5-fluorouracil and 5-fluorouracil with concurrent thymidine infusions in a phase I trial, *Cancer Res.*, 40, 107, 1980.

51. **Hsu, L. S. F. and Marrs, T. C.,** Determination of 5-fluorouracil in human plasma by high-pressure ion-exchange chromatography, *Ann. Clin. Biochem.*, 17, 272, 1980.
52. **Hsu, L. S. F.,** The kinetics of uptake of 5-fluorouracil by rat liver, *Br. J. Pharmacol.*, 77, 413, 1982.
53. **Cohen, J. L. and Brown, R. E.,** High-performance liquid chromatographic analysis of 5-fluorouracil in plasma, *J. Chromatogr.*, 151, 237, 1978.
54. **Washtein, W. and Santi, D. V.,** Assay of intracellular free and macromolecular-bound metabolites of 5-fluorodeoxyuridine and 5-fluorouracil, *Cancer Res.*, 39, 3397, 1979.
55. **Sitar, D. S., Shaw, D. H., Thirlwell, M. P., and Ruedy, J. R.,** Disposition of 5-fluorouracil after intravenous bolus doses of a commercial formulation to cancer patients, *Cancer Res.*, 37, 3981, 1977.
56. **Celio, L. A., DiGregorio, J., Ruch, E., Pace, J. N., and Piraino, A. J.,** 5-Fluorouracil concentrations in rat plasma, parotid saliva, and bile and protein binding in rat plasma, *J. Pharm. Sci.*, 72, 597, 1983.
57. **Celio, L. A., DiGregorio, G. J., Ruch, E., Pace, J., and Piraino, A. J.,** Doxorubicin and 5-fluorouracil plasma concentrations and detectability in parotid saliva, *Eur. J. Clin. Pharmacol.*, 24, 261, 1983.
58. **Au, J. L. S., Wientjes, M. G., Luccioni, C. M., and Rustum, Y. M.,** Reversed-phase ion-pair high-performance liquid chromatographic assay of 5-fluorouracil, 5′-deoxy-5-fluorouridine, their nucleosides, mono-, di-, and triphosphate nucleotides with a mixture of quaternary ammonium ions, *J. Chromatogr.*, 228, 245, 1982.
59. **Sommadossi, J.-P., Gewirtz, D. A., Diasio, R. B., Aubert, C., Cano, J.-P., and Goldman, I. D.,** Rapid catabolism of 5-fluorouracil in freshly isolated rat hepatocytes as analyzed by high performance liquid chromatography, *J. Biol. Chem.*, 257, 8171, 1982.
60. **Gelijkens, C. F. and De Leenheer, A. P.,** Reversed-phase liquid chromatography of 5-fluorouracil nucleosides and nucleotides in the presence of quaternary ammonium ions, *J. Chromatogr.*, 194, 305, 1980.
61. **King, M. E., Pfeifle, C. E., and Howell, S. B.,** Intraperitoneal cytosine arabinoside therapy in ovarian carcinoma, *J. Clin. Oncol.*, 2, 662, 1984.
62. **Ho, D. H. W. and Freireich, E. J.,** Clinical pharmacology of arabinosylcytosine, in *Handbook of Experimental Pharmacology: Antineoplastic and Immunosuppressive Agents,* Vol. 38, Part 2, Sartorelli, A. C. and Johns, D. G., Eds., Springer-Verlag, New York, 1975, chap. 43.
63. **Creasey, W. A.,** Arabinosylcytosine, in *Handbook of Experimental Pharmacology: Antineoplastic and Immunosuppressive Agents,* Vol. 38, Part 2, Sartorelli, A. C. and Johns, D. G., Eds., Springer-Verlag, New York, 1975, chap. 42.
64. **Cheng, Y. and Capizzi, R. L.,** Enzymology of cytosine arabinoside, *Med. Ped. Oncol. Suppl.*, 1, 27, 1982.
65. **Momparler, R. L.,** Biochemical pharmacology of cytosine arabinoside, in *Handbook of Experimental Pharmacology: Antineoplastic and Immunosuppressive Agents,* Vol. 38, Part 2, Sartorelli, A. C. and Johns, D. G., Eds., Springer-Verlag, New York, 1975, chap. 45.

65a. **Kufe, D. W. and Major, P. P.,** Studies on the mechanism of action of cytosine arabinoside, in *Handbook of Experimental Pharmacology: Antineoplastic and Immunosuppressive Agents,* Vol. 38, Part 2, Sartorelli, A. C. and Johns, D. G., Eds., Springer-Verlag, New York, 1975, chap. 49.

65b. **Fridland, A.,** Mode of action of 1-β-D-arabinofuranosylcytosine in human lymphoblasts: differential effects of the drug on chromosomal replication, in *Handbook of Experimental Pharmacology: Antineoplastic and Immunosuppressive Agents,* Vol. 38, Part 2, Sartorelli, A. C. and Johns, D. G., Eds., Springer-Verlag, New York, 1975, chap. 69.

66. **Ho, D. H. W. and Frei, E., III,** Clinical pharmacology of 1-β-D-arabinofuranosyl cytosine, *Clin. Pharmacol. Ther.*, 12, 944, 1971.
67. **Dollinger, M. R., Burchenal, J. H., Kreis, W., and Fox, J. J.,** Analogs of 1-β-D-arabinofuranosylcytosine: studies on mechanisms of action in Burkitts cell culture and mouse leukemia and *in vitro* deamination studies, *Biochem. Pharmacol.*, 16, 689, 1967.
68. **Chou, T.-C., Hutchison, D. J., Schmid, F. A., and Philips, F. S.,** Metabolism and selective effects of 1-β-D-arabinofuranosylcytosine in L1210 and host tissues in vivo, *Cancer Res.*, 35, 225, 1975.
69. **Mellett, L. B., Wyatt, P. J., and Woolley, C.,** Laser light scattering bioassay for 1-β-D-arabinofuranosylcytosine, *Res. Commun. Chem. Pathol. Pharmacol.*, 20, 379, 1978.
70. **Hanka, L. J., Kuentzel, S. L., and Neil, G. L.,** Improved microbiological assay for cytosine arabinoside (NSC-63878), *Cancer Chem. Rep.*, 54, 393, 1970.
71. **Mehta, B. M., Meyers, M. B., and Hutchison, D. J.,** Microbiologic assay for cytosine arabinoside (NSC-63878): the use of a mutant of *Streptococcus faecium* var. *durans* resistant to methotrexate (NSC-740) and 6-mercaptopurine (NSC-755), *Cancer Chem. Rep.*, 59, 515, 1975.
72. **van Prooijen, H. C., Vierwinden, G., van Egmond, J., Wessels, J. M. C., and Haanen, C.,** A sensitive bio-assay for pharmacokinetic studies of cytosine arabinoside in man, *Eur. J. Cancer*, 12, 899, 1976.
73. **Baguley, B. C. and Falkenhaug, E.-M.,** Plasma half-life of cytosine arabinoside (NSC-63878) in patients treated for acute myeloblastic leukemia, *Cancer Chem. Rep.*, 55, 291, 1971.

74. **Kreis, W., Greenspan, A., Woodcock, T., and Gordon, C.,** Separation of 5′-mono-, di-, and triphosphates of 1-β-D-arabinofuranosylcytosine and of 1-β-D-arabinofuranosyluracil by thin-layer and high pressure liquid chromatography, *J. Chromatogr. Sci.*, 14, 331, 1976.
75. **Momparler, R. L., Labitan, A., and Rossi, M.,** Enzymatic estimation and metabolism of 1-β-D-arabinofuranosylcytosine in man, *Cancer Res.*, 32, 408, 1972.
76. **Pantarotto, C., Martini, A., Belvedere, G., Bossi, A., Donelli, M. G., and Frigerio, A.,** Application of gas chromatography-chemical ionization mass fragmentography in the evaluation of bases and nucleoside analogues used in cancer chemotherapy, *J. Chromatogr.*, 99, 519, 1974.
77. **Harris, A. L., Potter, C., Bunch, C., Boutagy, J., Harvey, D. J., and Grahame-Smith, D. G.,** Pharmacokinetics of cytosine arabinoside in patients with acute myeloid leukaemia, *Br. J. Clin. Pharmacol.*, 8, 219, 1979.
78. **Boutagy, J. and Harvey, D. J.,** Determination of cytosine arabinoside in human plasma by gas chromatography with a nitrogen-sensitive detector and by gas chromatography-mass spectrometry, *J. Chromatogr.*, 146, 283, 1978.
79. **Piall, E. M., Aherne, G. W., and Marks, V. M.,** A radioimmunoassay for cytosine arabinoside, *Br. J. Cancer*, 40, 548, 1979.
80. **Okabayashi, T., Mihara, S., Repke, D. B., and Moffatt, J. G.,** A radioimmunoassay for 1-β-D-arabinofuranosylcytosine, *Cancer Res.*, 37, 619, 1977.
81. **Linssen, P., Drenthe-Schonk, A., Wessels, H., and Haanen, C.,** Determination of 1-β-D-arabinofuranosylcytosine and 1-β-D-arabinofuranosyluracil in human plasma by high-performance liquid chromatography, *J. Chromatogr.*, 223, 371, 1981.
82. **Breithaupt, H. and Schick, J.,** Determination of cytarabine and uracil arabinoside in human plasma and cerebrospinal fluid by high-performance liquid chromatography, *J. Chromatogr.*, 225, 99, 1981.
83. **Pallavicini, M. G. and Mazrimas, J. A.,** High-performance liquid chromatographic analysis of cytosine arabinoside and metabolites in biological samples, *J. Chromatogr.*, 183, 449, 1980.
84. **Zimm, S., Collins, J. M., Miser, J., Chaterji, D., and Poplack, D. G.,** Cytosine arabinoside cerebrospinal fluid kinetics, *Clin. Pharmacol. Ther.*, 35, 826, 1984.
85. **Rustum, Y. M.,** High-pressure liquid chromatography: quantitative separation of purine and pyrimidine nucleosides and bases, *Anal. Biochem.*, 90, 289, 1978.
86. **Rustum, Y. M., Slocum, H. K., Wang, G., Bakshi, D., Kelly, E., Buscaglia, D., Wrzosek, C., Early, A. P., and Priesler, H.,** Relationship between plasma ara-C and intracellular ara-CTP pools under conditions of continuous infusion and high dose ara-C treatment, *Med. Ped. Oncol., Suppl.*, 1, 33, 1982.
87. **Kreis, W., Gordon, C., Gizoni, C., and Woodcock, T.,** Extraction and analytic procedures for cytosine arabinoside and 1-β-D-arabinofuranosyluracil and their 5′-mono-, di-, and triphosphates, *Cancer Treat. Rep.*, 61, 643, 1977.
88. **Bury, R. W. and Keary, P. J.,** Determination of cytosine arabinoside in human plasma by high-pressure liquid chromatography, *J. Chromatogr.*, 146, 350, 1978.
89. **Wan, S. H., Huffman, D. H., Azarnoff, D. L., Hoogstraten, B., and Larsen, W. E.,** Pharmacokinetic of 1-β-D-arabinofuranosylcytosine in humans, *Cancer Res.*, 34, 392, 1974.
90. **Murphy, M. P., Andrews, P. A., and Howell, S. B.,** unpublished observations.
91. **Sinkule, J. A. and Evans, W. E.,** High-performance liquid chromatographic assay for cytosine arabinoside, uracil arabinoside and some related nucleosides, *J. Chromatogr.*, 274, 87, 1983.
92. **Plunkett, W., Chubb, S., and Barlogie, B.,** Simultaneous determination of 1-β-D-arabinofuranosylcytosine 5′-triphosphate and 3-deazauridine 5′-triphosphate in human leukemia cells by high-performance liquid chromatography, *J. Chromatogr.*, 221, 425, 1980.
93. **Plunkett, W., Hug, V., Keating, M. J., and Chubb, S.,** Quantitation of 1-β-D-arabinofuranosylcytosine 5′-triphosphate in the leukemic cells from bone marrow and peripheral blood of patients receiving 1-β-D-arabinofuranosylcytosine therapy, *Cancer Res.*, 40, 588, 1980.
94. **Danks, M. K.,** Two simple high-performance liquid chromatographic methods for simultaneous determination of 2′-deoxycytidine 5′-triphosphate and cytosine arabinoside 5′-triphosphate concentrations in biological samples, *J. Chromatogr.*, 233, 141, 1982.
95. **Rustum, Y. and Preisler, H. D.,** Correlation between leukemic cell retention of 1-β-D-arabinofuranosylcytosine 5′-triphosphate and response to therapy, *Cancer Res.*, 39, 42, 1979.
96. **Linssen, P., Drenthe-Schonk, A., Wessels, H., Vierwinden, G., and Haanen, C.,** Determination of cytosine arabinoside triphosphate in leukemic cells by isocratic high-performance anion-exchange column chromatography, *J. Chromatogr.*, 232, 424, 1982.

HPLC INVESTIGATION OF PT(II) COORDINATION COMPLEXES WITH NUCLEIC ACID COMPONENTS

B. Wenclawiak and W. Kleiböhmer

Since the discovery of the chemotherapeutic activity of *cis*-$Pt(NH_3)_2Cl_2$ (*cis*-diamminedichloroplatinum [II]) (cisDDP) by Rosenberg and co-workers in 1965,[1] reactions of *cis*-coordinated platinum compounds with RNA, DNA fragments and their building units have been studied intensively. All platinum compounds which have been used so far are reacting via their aquo-complexes $(Pt[ligand]_{1\text{-}2}[H_2O]_2)^{2+}$,[2,3] even if they contain different leaving groups like chloro-, malonato- or oxalato-ligands, at the N-coordination sites of the nucleobases guanine (N7), adenosine (N7 and N1), cytosine (N3 and NH_2), and thymine (N3,0).[4] Monofunctional coordination to one nucleobase or bifunctional coordination to one or two nucleobases (inter- or intrastrand cross-link to the DNA double helix) is possible.[5,6]

In order to get a better understanding of the mode of action of platinum drugs, many reactions between different platinum compounds and different DNA-, RNA-building units have been investigated. The reaction products have been characterized by different techniques like nuclear magnetic resonance spectroscopy (NMR), circular dichroism (CD), infrared spectroscopy (IR), ultraviolet spectroscopy (UV), X-ray crystal structure analysis, and high-performance liquid chromatography (HPLC) with UV detection. The amount of platinum in the products has commonly been determined by electrothermal atomic absorption spectroscopy (ETAAS), and the ratio of the Pt/nucleobase, nucleoside, or nucleotide was calculated by peak height comparison of the freed nucleounits in a chromatogram.

Up to now, the chromatography of *cis*-platinum compounds coordinated to DNA fragments has been performed using three different methods: (1) gel-filtration (GP),[7-10] (2) ion exchange (IE),[11-13,23] and reversed-phase chromatography (RPLC).[7-10, 14-17] The RP/C18 packed columns have also been solvent generated to ion exchange behavior.[18-22]

Gel chromatography (G-25, G-10 Sephadex) was used for the separation and desalting of the reaction products of $(Pt[NH_3]_2[H_2O]_2[NO_3]_2)$ with dinucleotides (IpI, GpG, ApA, GpC, dGpG, and dpGpG),[7] ribose-, deoxyribosediguanosinephosphates,[8] CpG, dpCpG, dpGpC, CpC, and dpCpC,[10] and deoxyhexanucleotide d(T-G-G-C-C-A).[9]

IEC of *cis*-DDP reaction products of dG, dA, dC, and thymidine was performed on strong cation-exchange resins.[11] Investigations with adenosine/*cis*-DDP and $(Pt[NH_3]_3Cl)Cl$ reactions at different mole ratios have also been performed.[12] Aminex® A6 cation exchange was used for time-course studies on the reaction system *cis*-DDP/dG.[13] Anion-exchange chromatography on DEAE Sephacel® was used for the separation of *cis*-$(Pt[NH_3]_2d[pGpG])$ and Pt-adducts of pGpXpG, with X = A, G, T, C.[23,24]

The solvent-generated anion-exchange chromatography has been widely used for the separation of inorganic platinum compounds[19,20] in urine[21] and for monitoring the *cis*-platinum/DNA nucleotide reaction.[18] Ion-pair reversed-phase chromatography (IP-RPLC) of *cis*-platinum/methionine was reported in the same paper.

RPLC is becoming a very useful tool for separating platinum adducts containing organic ligands. Using this technique, *cis*-$Pt(NH_3)_2$(thymine-H$)_2$, with one coordination to each thymine either at N1 or N3,[14] was observed. Both isocratic[7-9] and gradient elution chromatography have been used to characterize platinum nucleic acid adducts (see Table 1).[14-17] Gradient elution RPLC was first applied to the separation of *cis*-DDP/dG, dA, and dC reaction mixtures by Eastman,[15] who investigated the reaction products of different Pt/nucleoside ratios and adducts formed with DNA and *cis*-DDP and *cis*-DEP (*cis*-dichloroethylenediamineplatinum [II]). The interactions between *cis*-DDP and guanine or methylated guanines have been investigated by Woollins and Rosenberg[25] using HPLC. A

Table 1
HPLC INVESTIGATION ON THE INTERACTION OF PLATINUM COMPOUNDS AND NUCLEOACID FRAGMENTS AS CITED IN THIS PAPER

	Ref.
Nucleobases and derivatives	25
RNA fragments	
Nucleosides	12, 17
Adenosine, guanosine	
Dinucleotides	7, 8, 10, 17,
ApA, ApG, GpA, GpG, ApC, CpC, IpI	this paper
DNA fragments	
Nucleosides	11, 13, 15
dA, dC, dG	
Dinucleotides	8, 10, 23
d-pCpG, d-pGpC, d-pGpG, d-GpG	
Polynucleotides	9, 16
d-(CTpGpGpCpCpA), poly dG, poly dC, poly dGdC, poly dGdT, poly dAdC, poly dA × poly dT, polydAdT × poly dAdT	

time-course study of three different *cis*-platinum compounds (*cis*-DDP, *cis*-DEP, and *cis*-DPP = *cis*-dichloropropylenediamineplatinum [II]), reacted with RNA fragments such as adenosine and guanosine, was investigated by gradient RPLC.[17] Since the nucleic acid bases are multisite ligands, a number of product peaks were found. The chromatograms of *cis*-DDP and *cis*-DEP reacted with the dinucleotides showed the largest number of peaks for ApA. Since guanine prefers the N7 coordination site, the substitution of adenine by guanine changed the variety of products. Only one main product was formed with GpG.[17]

No investigations have been made, so far, on time-dependent formation of dinucleotides such as GpG, GpA, ApG, and ApA with *cis*-DDP in 1-m*M* ratios. The chromatographic system (RP/C18, 5 μm, NH_4OAc buffer 0.1 *M*, pH 4.25, gradient MeOH 0 to 30% within 30 min) and the reaction conditions (37°C, dark vials, stirring the mixture) have been described earlier.[17] During the time course, an aliquot of 20 μℓ was injected every hour and either peak heights or peak areas have been used to measure the reaction rate. The reaction was stopped after 24 hr. The formation of GpG-/ApG-*cis*-DDP was complete (less than 5% of starting dinucleotide was left), while ApA-*cis*-DDP still contained an important amount of ApA after 56 hr of reaction. GpG gave only one major adduct (Figure 1), while ApA gave two main adducts (Figure 2), which are eluted before the unreacted dinucleotides (Table 2). ApA and GpA formed intermediates which have not been observed before. The chromatograms showed one major peak and few minor peaks in the ApG-*cis*-DDP system, while GpA-*cis*-DDP resulted in a number of medium size and minor peaks after 24 hr of reaction.

Semipreparative chromatography with the same stationary phase was carried out for peak isolation. The comparison of the chromatograms of the analytical and semipreparative systems showed that the latter method resulted in a better detectability of minor peaks (785 μℓ injection volume instead of 20 μℓ). We have collected all the major peaks and a representative minor peak of the GpA mixture. The existence of Pt in all product peaks of the dinucleotide/*cis*-DDP systems was proved in the following manner:

1. A red shift of maximum absorbance recorded at stopped flow at peak maxima compared with pure dinucleotide measurements proved the existence of Pt-coordination (Table 2).

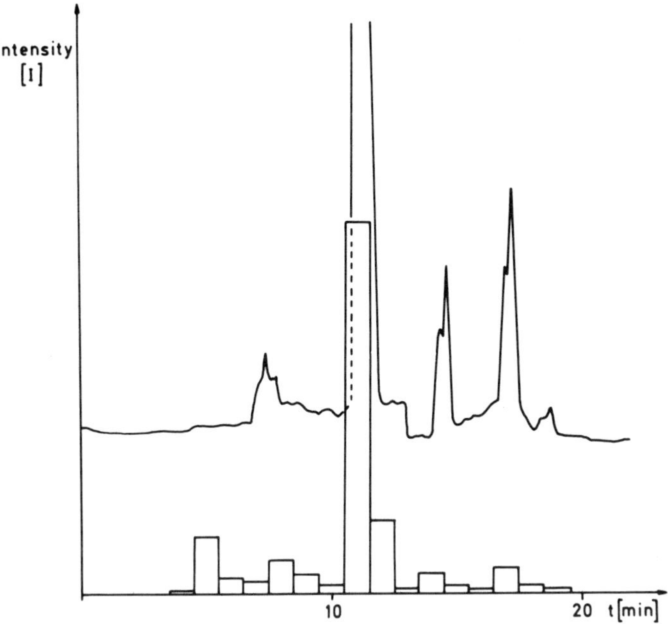

FIGURE 1. HPLC chromatogram of the reaction *cis*-DDP/GpG after 24 hr at 37°C while stirring, and platinum determination by graphite AAS of each fraction. The chromatogram using the analytical column does not show the last two peaks. Chromatographic conditions: column: RP/C18 250 × 8.0 mm ID; mobile phase: (A) 0.1 *M* NH_4OAc buffer, pH 4.25, (B) CH_3OH, linear gradient from 100 to 70% A in 15 min; flow rate: 2.0 mℓ/min; temperature: ambient; detection: UV detection, 254 nm.

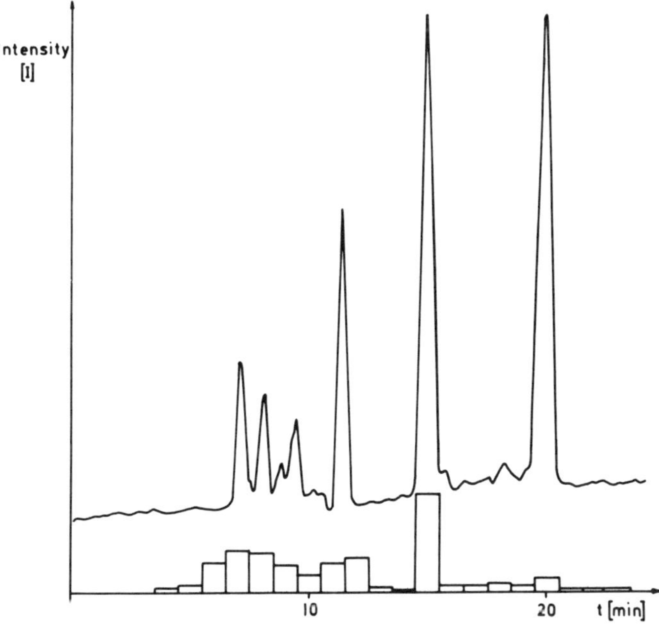

FIGURE 2. HPLC chromatogram of the reaction *cis*-DDP/ApA after 56 hr at 37°C while stirring, and platinum determination by ETAAS (1 cm height = 230 μg Pt in the fraction) of the collected fractions. Chromatographic conditions are the same as in Figure 1.

Table 2
RETENTION TIMES AND UV MAXIMA OF PURE AND PLATINUM COORDINATED MAJOR PEAKS

	t_r (min)	UV_{max} (nm)
	Pure	
ApA	36.6	257
ApG	34.2	257
GpA	32.6	257
GpG	27.5	257
	Pt-Adducts	
ApA I	27.5	264
ApA II	20.3	264
ApG	24.6	262
GpA I	10.8	—
GpA II	16.4	265
GpA III	25.4	—
GpG	18.5	260

Note: Retention times are taken from chromatograms of mixtures having reacted for 10 hr.

2. The total elution volume was divided into 25 fractions, each containing 2 mℓ. The amount of platinum in every fraction was determined by ETAAS (Figures 1 and 2).

Further studies are necessary to determine the exact ratio of Pt/dinucleotide of the different adducts.

ABBREVIATIONS

For the nucleotides, IUPAC/IUB recommendations are followed (see *Eur. J. Biochem.*, 15, 203, 1970.

REFERENCES

1. **Rosenberg, B., van Camp, L., and Krigas, T.,** Binding of platinum compounds to nucleic acids with respect to the anti-tumor activity of cis-diamminedichloroplatinum (II) and derivatives, *Nature,* 205, 698, 1965.
2. **Horacek, P. and Drobnik, J.,** Interaction of cis-dichlorodiammineplatinum(II) with DNA, *Biochim. Biophys. Acta,* 254, 341, 1971.
3. **Drobnik, J. and Horacek, P.,** Specific biological activity of platinum complexes. Theory of molecular mechanism, *Chem.-Biol. Interactions,* 7(4), 223, 1973.
4. **Mansey, S., Rosenberg, B., and Thomson, H. J.,** Binding of cis- and trans-dichlorodiammineplatinum(II) to nucleosides. Location of the binding sites, *J. Am. Chem. Soc.,* 95, 1633, 1973.

5. **Pascoe, J. M. and Roberts, J. J.,** Interactions between mammalian cell DNA and inorganic platinum compounds. I. DNA interstrand crosslinking and cytotoxic properties of platinum(II) compounds, *Biochim. Pharmacol.*, 23, 1345, 1974.
6. **Stone, P. J., Kelman, A. D., and Sinex, F. M.,** Specific binding of antitumor drug-cis-diamminedichloroplatinum to DNA rich in guanine and cytosine, *Nature*, 251, 736, 1974.
7. **Chottard, J. C., Girault, J. P., Chottard, G., Lallemand, J. Y., and Mansuy, D.,** Interaction of cis-$[Pt(NH_3)_2(H_2O)_2](NO_3)_2$ with ribose dinucleotides monophosphates, *J. Am. Chem. Soc.*, 102, 5565, 1980.
8. **Girault, J. C., Chottard, G., Lallemand, J. Y., and Chottard, J. C.,** Interaction of cis-$[Pt(NH_3)_2(H_2O)_2](NO_3)_2$ with ribose and desoxyribose diguanosine phosphates, *Biochemistry*, 21, 1352, 1982.
9. **Girault, J. P., Chottard, J. C., Guittet, E. R., Lallemand, J. Y., Huynh-Dinh, T., and Igolen, J.,** Specific platinum chelation by the guanines of the desoxyhexanucleotide d(I-G-G-C-C-A) upon reaction with cis-$[Pt(NH_3)_2(H_2O)_2](NO_3)_2$, *Biochem. Biophys. Res. Commun.*, 109(4), 1157, 1982.
10. **Chottard, J. C., Girault, J. P., Guittet, E. R., Lallemand, J. Y., and Chottard, G.,** Platinum-oligonucleotide structures and their relevance to platinum-DNA interaction, in *Platinum, Gold and Other Metal Chemotherapeutic Agents*, ACS Symp. Ser. 209, Lippard, S. J., Ed., Washington, D.C., 1983, 125.
11. **Inagaki, K., Tamaoki, N., and Kidani, Y.,** Studies on reactivity of deoxynucleosides with cis-$Pt(NH_3)_2Cl_2$ and the related complexes by high performance liquid chromatography, *Inorg. Chim. Acta*, 46(5), L93, 1980.
12. **Inagaki, K., Kuwayama, M., and Kidani, J.,** Studies of the reaction products of adenosine with $[Pt(NH_3)_3Cl]Cl$ and cis-$Pt(NH_3)Cl_2$ by high performance liquid chromatography, *J. Inorg. Biochem.*, 16(1), 59, 1982.
13. **Rahn, R. O., Chang, S. S., and Hoeschele, J. D.,** Adducts formed between desoxyguanosine and cis-dichlorodiammineplatinum(II)-: separation by means of cation-exchange chromatography, *J. Inorg. Biochem.*, 18(4), 2790, 1983.
14. **Woollins, J. D. and Rosenberg, B.,** High performance liquid chromatography studies on platinum thymine blue, *Inorg. Chem.*, 21, 1280, 1982.
15. **Eastman, A.,** Separation and characterization of products resulting from the reaction of cis-diamminedichloroplatinum(II) with desoxyribonucleotides, *Biochemistry*, 21, 6732, 1982.
16. **Eastman, A.,** Characterization of the adducts produced in DNA by cis-diamminedichloroplatinum(II) and cis-dichloro(ethylenediamine) platinum(II), *Biochemistry*, 22, 3927, 1983.
17. **Wenclawiak, B., Kleiböhmer, W., and Krebs, B.,** High performance liquid chromtographic investigations on the time dependent reaction of cis-$Pt(NH_3)_2Cl_2$, $Pt(en)Cl_2$ and $Pt(pn)Cl_2$ with RNA fragments, *J. Chromatogr.*, 296, 395, 1984.
18. **Riley, C. M., Sternson, L. A., Repta, A. J., and Slyter, S. A.,** Monitoring the reactions of cisplatin with nucleotides and methionine by reversed-phase high-performance liquid chromatography using cationic and anionic paring ions, *Anal. Biochem.*, 130(1), 203, 1983.
19. **Riley, C. M., Sternson, L. A., and Repta, A. J.,** High-performance liquid chromatography of cis-dichlorodiammineplatinum(II) using chemically-bonded and solvent generated ion exchangers, *J. Chromatogr.*, 217, 405, 1981.
20. **Riley, C. M., Sternson, L. A., and Repta, A. J.,** High-performance liquid chromatography of inorganic platinum(II) complexes using solvent generated anion exchangers. II. The effect of electrolytes on solute retention, *J. Chromatogr.*, 219(2), 235, 1981.
21. **Riley, C. M., Sternson, L. A., Repta, A. J., and Siegler, R. W.,** High-performance liquid chromatography of platinum complexes on solvent generated anion exchangers. III. Application to the analysis of cisplatin in urine using automated column switching, *J. Chromatogr.*, 229(2), 373, 1982.
22. **Riley, C. M., Sternson, L. A., and Repta, A. J.,** Assessment of cisplatin reactivity with peptides and proteins using reverse phase high performance liquid chromatography and flameless atomic absorption spectroscopy, *Anal. Biochem.*, 124(1), 167, 1982.
23. **Fichtinger-Schepmann, A. M. J., Lohmann, P. H. M., and Reedijk, J.,** Detection and quantification of adducts formed upon interaction of diamminedichloroplatinum(II) with DNA by anion-exchange chromatography after enzymic degradation, *Nucleic Acids Res.*, 10(17), 5346, 1983.
24. **Marcelis, A. T. M. and Reedijk, J.,** Binding of platinum compounds to nucleic acids with respect to the anti-tumor activity of cis-diamminedichloroplatinum (II) and derivatives, *Recl. Trav. Chim. Pays-Bas.*, 102, 121, 1983.
25. **Woollins, A. and Rosenberg, B.,** High-performance liquid chromatography studies on the interactions of cis-diamminediaquoplatinumion (cis-$Pt(NH_3)_2(H_2O)_2^{2+}$) with guanine and methylated guanines, *J. Inorg. Biochem.*, 20(1), 23, 1984.

Index

INDEX

C

D

E

F

G

H

V

W

X

Y

Z